NISTIR 7628

Guidelines for Smart Grid Cyber Security: Vol. 1, Smart Grid Cyber Security Strategy, Architecture, and High-Level Requirements

The Smart Grid Interoperability Panel–Cyber Security Working Group

August 2010

U. S. Department of Commerce
Gary Locke, Secretary

National Institute of Standards and Technology
Patrick D. Gallagher, Director

REPORTS ON COMPUTER SYSTEMS TECHNOLOGY

The Information Technology Laboratory (ITL) at the National Institute of Standards and Technology (NIST) promotes the U.S. economy and public welfare by providing technical leadership for the Nation's measurement and standards infrastructure. ITL develops tests, test methods, reference data, proof of concept implementations, and technical analysis to advance the development and productive use of information technology (IT). ITL's responsibilities include the development of technical, physical, administrative, and management standards and guidelines for the cost-effective security and privacy of sensitive unclassified information in federal computer systems. This National Institute of Standards and Technology Interagency Report (NISTIR) discusses ITL's research, guidance, and outreach efforts in computer security and its collaborative activities with industry, government, and academic organizations.

> Certain commercial entities, equipment, or materials may be identified in this report in order to describe an experimental procedure or concept adequately. Such identification is not intended to imply recommendation or endorsement by the National Institute of Standards and Technology, nor is it intended to imply that the entities, materials, or equipment are necessarily the best available for the purpose.

ACKNOWLEDGMENTS

This report was developed by members of the Smart Grid Interoperability Panel–Cyber Security Working Group (SGIP-CSWG), formerly the Cyber Security Coordination Task Group (CSCTG), and during its development was chaired by Annabelle Lee of the Federal Energy Regulatory Commission (FERC), formerly of NIST. The CSWG is now chaired by Marianne Swanson (NIST). Alan Greenberg (Boeing), Dave Dalva (Cisco Systems), and Bill Hunteman (Department of Energy) are the vice chairs. Mark Enstrom (Neustar) is the secretary. Tanya Brewer of NIST is the lead editor of this report. The members of the SGIP-CSWG have extensive technical expertise and knowledge to address the cyber security needs of the Smart Grid. The dedication and commitment of all these individuals over the past year and a half is significant. In addition, appreciation is extended to the various organizations that have committed these resources to supporting this endeavor. Members of the SGIP-CSWG and the working groups of the SGIP-CSWG are listed in Appendix J of this report.

In addition, acknowledgement is extended to the NIST Smart Grid Team, consisting of staff in the NIST Smart Grid Office and several of NIST's Laboratories. Under the leadership of Dr. George Arnold, National Coordinator for Smart Grid Interoperability, their ongoing contribution and support of the CSWG efforts have been instrumental to the success of this report.

Additional thanks are extended to Diana Johnson (Boeing) and Liz Lennon (NIST) for their superb technical editing of this report. Their expertise, patience, and dedication were critical in producing a quality report. Thanks are also extended to Victoria Yan (Booz Allen Hamilton). Her enthusiasm and willingness to jump in with both feet are really appreciated.

Finally, acknowledgment is extended to all the other individuals who have contributed their time and knowledge to ensure this report addresses the security needs of the Smart Grid.

TABLE OF CONTENTS

EXECUTIVE SUMMARY VIII
    Content of the Report x
CHAPTER ONE   CYBER SECURITY STRATEGY 1
    1.1   Cyber Security and the Electric Sector 3
    1.2   Scope and Definitions 4
    1.3   Smart Grid Cyber Security Strategy 5
    1.4   Outstanding Issues and Remaining Tasks 12
CHAPTER TWO   LOGICAL ARCHITECTURE AND INTERFACES OF THE SMART GRID 14
    2.1   The Seven Domains to the Logical Reference Model 15
    2.2   Logical Security Architecture Overview 25
    2.3   Logical Interface Categories 26
CHAPTER THREE   HIGH-LEVEL SECURITY REQUIREMENTS 72
    3.1   Cyber Security Objectives 72
    3.2   Confidentiality, Integrity, and Availability Impact Levels 73
    3.3   Impact Levels for the CI&A Categories 74
    3.4   Selection of Security Requirements 76
    3.5   Security Requirements Example 77
    3.6   Recommended Security Requirements 78
    3.7   Access Control (SG.AC) 90
    3.8   Awareness and Training (SG.AT) 103
    3.9   Audit and Accountability (SG.AU) 107
    3.10 Security Assessment and Authorization (SG.CA) 116
    3.11 Configuration Management (SG.CM) 120
    3.12 Continuity of Operations (SG.CP) 127
    3.13 Identification and Authentication (SG.IA) 134
    3.14 Information and Document Management (SG.ID) 138
    3.15 Incident Response (SG.IR) 141
    3.16 Smart Grid Information System Development and Maintenance (SG.MA) 148
    3.17 Media Protection (SG.MP) 153
    3.18 Physical and Environmental Security (SG.PE) 156
    3.19 Planning (SG.PL) 163
    3.20 Security Program Management (SG.PM) 167
    3.21 Personnel Security (SG.PS) 171
    3.22 Risk Management and Assessment (SG.RA) 176
    3.23 Smart Grid Information System and Services Acquisition (SG.SA) 181
    3.24 Smart Grid Information System and Communication Protection (SG.SC) 187
    3.25 Smart Grid Information System and Information Integrity (SG.SI) 203
CHAPTER FOUR   CRYPTOGRAPHY AND KEY MANAGEMENT 210
    4.1   Smart Grid Cryptography and Key Management Issues 210
    4.2   Cryptography and Key Management Solutions and Design Considerations 219
    4.3   NISTIR High-Level Requirement Mappings 232
    4.4   References & Sources 252
APPENDIX A CROSSWALK OF CYBER SECURITY DOCUMENTS A-1
APPENDIX B EXAMPLE SECURITY TECHNOLOGIES AND SERVICES TO MEET THE HIGH-LEVEL SECURITY REQUIREMENTS B-1
    B.1   Power System Configurations and Engineering Strategies B-1

B.2 Local Equipment Monitoring, Analysis, and Control B-2
B.3 Centralized Monitoring and Control B-3
B.4 Centralized Power System Analysis and Control B-3
B.5 Testing B-4
B.6 Training B-4
B.7 Example Security Technology and Services B-4

LIST OF FIGURES

Figure 1-1 Tasks in the Smart Grid Cyber Security Strategy 7
Figure 2-1 Interaction of Actors in Different Smart Grid Domains through Secure Communication Flows 15
Figure 2-2 Composite High-level View of the Actors within Each of the Smart Grid Domains 16
Figure 2-3 Logical Reference Model 17
Figure 2-4 Logical Interface Category 1 33
Figure 2-5 Logical Interface Category 2 34
Figure 2-6 Logical Interface Category 3 35
Figure 2-7 Logical Interface Category 4 36
Figure 2-8 Logical Interface Category 5 38
Figure 2-9 Logical Interface Category 6 40
Figure 2-10 Logical Interface Category 7 42
Figure 2-11 Logical Interface Category 8 43
Figure 2-12 Logical Interface Category 9 45
Figure 2-13 Logical Interface Category 10 47
Figure 2-14 Logical Interface Category 11 48
Figure 2-15 Logical Interface Category 12 49
Figure 2-16 Logical Interface Category 13 51
Figure 2-17 Logical Interface Category 14 53
Figure 2-18 Logical Interface Category 15 56
Figure 2-19 Logical Interface Category 16 59
Figure 2-20 Logical Interface Category 17 62
Figure 2-21 Logical Interface Category 18 64
Figure 2-22 Logical Interface Category 19 65
Figure 2-23 Logical Interface Category 20 67
Figure 2-24 Logical Interface Category 21 69
Figure 2-25 Logical Interface Category 22 71

LIST OF TABLES

Table 1-1 Categories of Adversaries to Information Systems 9
Table 2-1 Actor Descriptions for the Logical Reference Model 18
Table 2-2 Logical Interfaces by Category 27
Table 3-1 Impact Levels Definitions 74
Table 3-2 Smart Grid Impact Levels 75

Table 3-3 Allocation of Security Requirements to Logical Interface Catgories 79
Table 4-1 Symmetric Key – Approved Algorithms 235
Table 4-2 Asymetric Key – Approved Algortihms 236
Table 4-3 Secure Hash Standard (SHS) – Approved Algorithms 237
Table 4-4 Message Authentication – Approved Algortihms 237
Table 4-5 Key Management – Approved Algortihms 238
Table 4-6 Deterministic Random Number Generators – Approved Algorithms 239
Table 4-7 Non-Deterministic Random Number Generators – Algorithms 240
Table 4-8 Symmetric Key Establishment Techniques – Approved Algortihms 241
Table 4-9 Asymmetric Key Establishment Techniques – Approved Algortihms 241
Table 4-10 Comparable Key Strengths 243
Table 6-11 Crypto Lifetimes 244
Table 4-12 Hash Function Security Strengths 245
Table 4-13 KMS Requirements 248
Table A-1 Crosswalk of Cyber Security Requirements and Documents A-1
Table B-2 Example Security Technologies and Services B-5

EXECUTIVE SUMMARY

The United States has embarked on a major transformation of its electric power infrastructure. This vast infrastructure upgrade—extending from homes and businesses to fossil-fuel-powered generating plants and wind farms, affecting nearly everyone and everything in between—is central to national efforts to increase energy efficiency, reliability, and security; to transition to renewable sources of energy; to reduce greenhouse gas emissions; and to build a sustainable economy that ensures future prosperity. These and other prospective benefits of "smart" electric power grids are being pursued across the globe.

Steps to transform the nation's aging electric power grid into an advanced, digital infrastructure with two-way capabilities for communicating information, controlling equipment, and distributing energy will take place over many years. In concert with these developments and the underpinning public and private investments, key enabling activities also must be accomplished. Chief among them is devising effective strategies for protecting the privacy of Smart Grid-related data and for securing the computing and communication networks that will be central to the performance and availability of the envisioned electric power infrastructure. While integrating information technologies is essential to building the Smart Grid and realizing its benefits, the same networked technologies add complexity and also introduce new interdependencies and vulnerabilities. Approaches to secure these technologies and to protect privacy must be designed and implemented early in the transition to the Smart Grid.

This three-volume report, *Guidelines for Smart Grid Cyber Security*, presents an analytical framework that organizations can use to develop effective cyber security strategies tailored to their particular combinations of Smart Grid-related characteristics, risks, and vulnerabilities. Organizations in the diverse community of Smart Grid stakeholders—from utilities to providers of energy management services to manufacturers of electric vehicles and charging stations—can use the methods and supporting information presented in this report as guidance for assessing risk and identifying and applying appropriate security requirements. This approach recognizes that the electric grid is changing from a relatively closed system to a complex, highly interconnected environment. Each organization's cyber security requirements should evolve as technology advances and as threats to grid security inevitably multiply and diversify.

This initial version of *Guidelines for Smart Grid Cyber Security* was developed as a consensus document by the Cyber Security Working Group (CSWG) of the Smart Grid Interoperability Panel (SGIP), a public-private partnership launched by the National Institute of Standards and Technology (NIST) in November 2009.[1] The CSWG now numbers more than 475 participants from the private sector (including vendors and service providers), manufacturers, various standards organizations, academia, regulatory organizations, and federal agencies. A number of these members are from outside of the U.S.

---

[1] For a brief overview of this organization, read the *Smart Grid Interoperability Panel: A New, Open Forum for Standards Collaboration* at: http://collaborate.nist.gov/twiki-sggrid/pub/SmartGrid/CMEWG/Whatis_SGIP_final.pdf.

This document is a companion document to the *NIST Framework and Roadmap for Smart Grid Interoperability Standards, Release 1.0* (NIST SP 1108),[2] which NIST issued on January 19, 2010. The framework and roadmap report describes a high-level conceptual reference model for the Smart Grid, identifies standards that are applicable (or likely to be applicable) to the ongoing development of an interoperable Smart Grid, and specifies a set of high-priority standards-related gaps and issues. Cyber security is recognized as a critical, cross-cutting issue that must be addressed in all standards developed for Smart Grid applications. Given the transcending importance of cyber security to Smart Grid performance and reliability, this document "drills down" from the initial release of the *NIST Framework and Roadmap*, providing the technical background and additional details that can inform organizations in their risk management efforts to securely implement Smart Grid technologies. The Framework document is the first installment in an ongoing standards and harmonization process. Ultimately, this process will deliver the hundreds of communication protocols, standard interfaces, and other widely accepted and adopted technical specifications necessary to build an advanced, secure electric power grid with two-way communication and control capabilities. The *Guidelines for Smart Grid Cyber Security* expands upon the discussion of cyber security included in the Framework document. The CSWG will continue to provide additional guidance as the Framework document is updated and expanded to address testing and certification, the development of an overall architecture, and as additional standards are identified.

This document is the product of a participatory public process that, starting in March 2009, included workshops as well as weekly teleconferences, all of which were open to all interested parties. Drafts of the three volumes have undergone at least one round of formal public review. Portions of the document have undergone two rounds of review and comment, both announced through notices in the Federal Register.[3]

The three volumes that make up this initial set of guidelines are intended primarily for individuals and organizations responsible for addressing cyber security for Smart Grid systems and the constituent subsystems of hardware and software components. Given the widespread and growing importance of the electric infrastructure in the U.S. economy, these individuals and organizations comprise a large and diverse group. It includes vendors of energy information and management services, equipment manufacturers, utilities, system operators, regulators, researchers, and network specialists. In addition, the guidelines have been drafted to incorporate the perspectives of three primary industries converging on opportunities enabled by the emerging Smart Grid—utilities and other business in the electric power sector, the information technology industry, and the telecommunications sector.

Following this executive summary, the first volume of the report describes the analytical approach, including the risk assessment process, used to identify high-level security requirements. It also presents a high-level architecture followed by a logical interface

---

[2] Office of the National Coordinator for Smart Grid Interoperability, National Institute of Standards and Technology, *NIST Framework and Roadmap for Smart Grid Interoperability Standards, Release 1.0 (NIST SP 1108)*, Jan. 2010. The report can be downloaded at: http://nist.gov/smartgrid/

[3] 1) *Federal Register*: October 9, 2009 (Volume 74, Number 195) [Notices], pp. 52183-52184; 2) *Federal Register*: April 13, 2010 (Volume 75, Number 70) [Notices], pp. 18819-18823.

architecture used to identify and define categories of interfaces within and across the seven Smart Grid domains. High-level security requirements for each of the 22 logical interface categories are then described. The first volume concludes with a discussion of technical cryptographic and key management issues across the scope of Smart Grid systems and devices.

The second volume is focused on privacy issues within personal dwellings. It provides awareness and discussion of such topics as evolving Smart Grid technologies and associated new types of information related to individuals, groups of individuals, and their behavior within their premises and electric vehicles; and whether these new types of information may contain privacy risks and challenges that have not been legally tested yet. Additionally, the second volume provides recommendations, based on widely accepted privacy principles, for entities that participate within the Smart Grid. These recommendations include things such as having entities develop privacy use cases that track data flows containing personal information in order to address and mitigate common privacy risks that exist within business processes within the Smart Grid; and to educate consumers and other individuals about the privacy risks within the Smart Grid and what they can do to mitigate these risks.

The third volume is a compilation of supporting analyses and references used to develop the high-level security requirements and other tools and resources presented in the first two volumes. These include categories of vulnerabilities defined by the working group and a discussion of the bottom-up security analysis that it conducted while developing the guidelines. A separate chapter distills research and development themes that are meant to present paradigm changing directions in cyber security that will enable higher levels of reliability and security for the Smart Grid as it continues to become more technologically advanced. In addition, the third volume provides an overview of the process that the CSWG developed to assess whether standards, identified through the NIST-led process in support of Smart Grid interoperability, satisfy the high-level security requirements included in this report.

Beyond this executive summary, it is assumed that readers of this report have a functional knowledge of the electric power grid and a functional understanding of cyber security.

CONTENT OF THE REPORT

- Volume 1 – Smart Grid Cyber Security Strategy, Architecture, and High-Level Requirements

    - Chapter 1 – *Cyber Security Strategy* includes background information on the Smart Grid and the importance of cyber security in ensuring the reliability of the grid and the confidentiality of specific information. It also discusses the cyber security strategy for the Smart Grid and the specific tasks within this strategy.

    - Chapter 2 – *Logical Architecture* includes a high level diagram that depicts a composite high level view of the actors within each of the Smart Grid domains and includes an overall logical reference model of the Smart Grid, including all the major domains. The chapter also includes individual diagrams for each of the 22 logical interface categories. This architecture focuses on a short-term view (1–3 years) of the Smart Grid.

- Chapter 3 – *High Level Security Requirements* specifies the high level security requirements for the Smart Grid for each of the 22 logical interface categories included in Chapter 2.
- Chapter 4 – *Cryptography and Key Management* identifies technical cryptographic and key management issues across the scope of systems and devices found in the Smart Grid along with potential alternatives.
- Appendix A – *Crosswalk of Cyber Security Documents*
- Appendix B – *Example Security Technologies and Procedures to Meet the High Level Security Requirements*

- Volume 2 – Privacy and the Smart Grid
  - Chapter 5 – *Privacy and the Smart Grid* includes a privacy impact assessment for the Smart Grid with a discussion of mitigating factors. The chapter also identifies potential privacy issues that may occur as new capabilities are included in the Smart Grid.
  - Appendix C – *State Laws – Smart Grid and Electricity Delivery*
  - Appendix D – *Privacy Use Cases*
  - Appendix E – *Privacy Related Definitions*

- Volume 3 – Supportive Analyses and References
  - Chapter 6 – *Vulnerability Classes* includes classes of potential vulnerabilities for the Smart Grid. Individual vulnerabilities are classified by category.
  - Chapter 7 – *Bottom-Up Security Analysis of the Smart Grid* identifies a number of specific security problems in the Smart Grid. Currently, these security problems do not have specific solutions.
  - Chapter 8 – *Research and Development Themes for Cyber Security in the Smart Grid* includes R&D themes that identify where the state of the art falls short of meeting the envisioned functional, reliability, and scalability requirements of the Smart Grid.
  - Chapter 9 – *Overview of the Standards Review* includes an overview of the process that is being used to assess standards against the high level security requirements included in this report.
  - Chapter 10 – *Key Power System Use Cases for Security Requirements* identifies key use cases that are architecturally significant with respect to security requirements for the Smart Grid.
  - Appendix F – *Logical Architecture and Interfaces of the Smart Grid*
  - Appendix G – *Analysis Matrix of Interface Categories*
  - Appendix H – *Mappings to the High Level Security Requirements*
  - Appendix I – *Glossary and Acronyms*
  - Appendix J – *SGIP-CSWG Membership*

CHAPTER ONE
CYBER SECURITY STRATEGY

With the implementation of the Smart Grid has come an increase in the importance of the information technology (IT) and telecommunications infrastructures in ensuring the reliability and security of the electric sector. Therefore, the security of systems and information in the IT and telecommunications infrastructures must be addressed by an evolving electric sector. Security must be included in all phases of the system development life cycle, from design phase through implementation, maintenance, and disposition/sunset.

Cyber security must address not only deliberate attacks launched by disgruntled employees, agents of industrial espionage, and terrorists, but also inadvertent compromises of the information infrastructure due to user errors, equipment failures, and natural disasters. Vulnerabilities might allow an attacker to penetrate a network, gain access to control software, and alter load conditions to destabilize the grid in unpredictable ways. The need to address potential vulnerabilities has been acknowledged across the federal government, including the National Institute of Standards and Technology (NIST)[4], the Department of Homeland Security (DHS),[5] the Department of Energy (DOE),[6] and the Federal Energy Regulatory Commission (FERC).[7]

Additional risks to the grid include:

- Increasing the complexity of the grid could introduce vulnerabilities and increase exposure to potential attackers and unintentional errors;

- Interconnected networks can introduce common vulnerabilities;

- Increasing vulnerabilities to communication disruptions and the introduction of malicious software/firmware or compromised hardware could result in denial of service (DoS) or other malicious attacks;

- Increased number of entry points and paths are available for potential adversaries to exploit;

- Interconnected systems can increase the amount of private information exposed and increase the risk when data is aggregated;

- Increased use of new technologies can introduce new vulnerabilities; and

---

[4] Testimony of Cita M. Furlani, Director, Information Technology Laboratory, NIST, before the United States House of Representatives Homeland Security Subcommittee on Emerging Threats, Cyber security, and Science and Technology, March 24, 2009.

[5] Statement for the Record, Sean P. McGurk, Director, Control Systems Security Program, National Cyber Security Division, National Protection and Programs Directorate, Department of Homeland Security, before the U.S. House of Representatives Homeland Security Subcommittee on Emerging Threats, Cybersecurity, and Science and Technology, March 24, 2009.

[6] U.S. Department of Energy, Office of Electricity Delivery and Energy Reliability, Smart Grid Investment Grant Program, Funding Opportunity: DE-FOA-0000058, Electricity Delivery and Energy Reliability Research, Development and Analysis, June 25, 2009.

[7] Federal Energy Regulatory Commission, Smart Grid Policy, 128 FERC ¶ 61,060 [Docket No. PL09-4-000] July 16, 2009.

- Expansion of the amount of data that will be collected that can lead to the potential for compromise of data confidentiality, including the breach of customer privacy.

With the ongoing transition to the Smart Grid, the IT and telecommunication sectors will be more directly involved. These sectors have existing cyber security standards to address vulnerabilities and assessment programs to identify known vulnerabilities in their systems. These same vulnerabilities need to be assessed in the context of the Smart Grid infrastructure. In addition, the Smart Grid will have additional vulnerabilities due not only to its complexity, but also because of its large number of stakeholders and highly time-sensitive operational requirements.

In its broadest sense, cyber security for the power industry covers all issues involving automation and communications that affect the operation of electric power systems and the functioning of the utilities that manage them and the business processes that support the customer base. In the power industry, the focus has been on implementing equipment that can improve power system reliability. Until recently, communications and IT equipment were typically seen as supporting power system reliability. However, increasingly these sectors are becoming more critical to the reliability of the power system. For example, in the August 14, 2003, blackout, a contributing factor was issues with communications latency in control systems. With the exception of the initial power equipment problems, the ongoing and cascading failures were primarily due to problems in providing the right information to the right individuals within the right time period. Also, the IT infrastructure failures were not due to any terrorist or Internet hacker attack; the failures were caused by inadvertent events—mistakes, lack of key alarms, and poor design. Therefore, inadvertent compromises must also be addressed, and the focus must be an all-hazards approach.

Development of the *Guidelines for Smart Grid Cyber Security* began with the establishment of a Cyber Security Coordination Task Group (CSCTG) in March 2009 that was established and is led by the National Institute of Standards and Technology (NIST). The CSCTG now numbers more than 475 participants from the private sector (including vendors and service providers), manufacturers, various standards organizations, academia, regulatory organizations, and federal agencies. This group was renamed under the Smart Grid Interoperability Panel (SGIP) as the Cyber Security Working Group (SGIP-CSWG) in January 2010 (hereafter referred to as the CSWG).

Cyber security is being addressed using a thorough process that results in a high-level set of cyber security requirements. As explained more fully later in this chapter, these requirements were developed (or augmented, where standards/guidelines already exist) using a high-level risk assessment process that is defined in the cyber security strategy section of this report. Cyber security requirements are implicitly recognized as critical in all of the priority action plans discussed in the Special Publication (SP), *NIST Framework and Roadmap for Smart Grid Interoperability Standards,* Release 1.0 (NIST SP 1108), which was published in January 2010.[8]

The Framework document describes a high-level reference model for the Smart Grid, identifies 75 existing standards that can be used now to support Smart Grid development, identifies 15 high-priority gaps and harmonization issues (in addition to cyber security) for which new or

---

[8] Available at http://www.nist.gov/public_affairs/releases/upload/smartgrid_interoperability_final.pdf.

revised standards and requirements are needed, documents action plans with aggressive timelines by which designated standards-setting organizations (SSOs) are tasked to fill these gaps, and describes the strategy to establish requirements and standards to help ensure Smart Grid cyber security. This Framework document is the first installment in an ongoing standards and harmonization process. Ultimately, this process will deliver the hundreds of communication protocols, standard interfaces, and other widely accepted and adopted technical specifications necessary to build an advanced, secure electric power grid with two-way communication and control capabilities. The NISTIR expands upon the discussion of cyber security included in the Framework document. The NISTIR is a starting point and a foundation. CSWG will continue to provide additional guidance as the Framework document is updated and expanded to address testing and certification, the development of an overall architecture, and as additional standards are identified.

The CSWG has liaisons to other Smart Grid industry groups to support and encourage coordination among the various efforts. The documented liaisons are listed at http://collaborate.nist.gov/twiki-sggrdi/bin/view/SmartGrid/CSWGLiaisonInformation.

This report is a tool for organizations that are researching, designing, developing, and implementing Smart Grid technologies. The cyber security strategy, risk assessment process, and security requirements included in this report should be applied to the entire Smart Grid system.

Cyber security risks must be addressed as organizations implement and maintain their Smart Grid systems. Therefore, this report may be used as a guideline to evaluate the overall cyber risks to a Smart Grid system during the design phase and during system implementation and maintenance. The Smart Grid risk mitigation strategy approach defined by an organization will need to address the constantly evolving cyber risk environment. The goal is to identify and mitigate cyber risk for a Smart Grid system using a risk methodology applied at the organization and system level, including cyber risks for specific components within the system. This methodology in conjunction with the system-level architecture will allow organizations to implement a Smart Grid solution that is secure and meets the reliability requirements of the electric grid.

The information included in this report is guidance for organizations. NIST is not prescribing particular solutions through the guidance contained in this report. Each organization must develop its own detailed cyber security approach (including a risk assessment methodology) for securing the Smart Grid.

1.1 CYBER SECURITY AND THE ELECTRIC SECTOR

The critical role of cyber security in ensuring the effective operation of the Smart Grid is documented in legislation and in the DOE Energy Sector Plan.

Section 1301 of the Energy Independence and Security Act of 2007 (P.L. 110-140) states:

> It is the policy of the United States to support the modernization of the Nation's electricity transmission and distribution system to maintain a reliable and secure electricity infrastructure that can meet future demand growth and to achieve each of the following, which together characterize a Smart Grid:
>
> (1) Increased use of digital information and controls technology to improve reliability, security, and efficiency of the electric grid.

(2) Dynamic optimization of grid operations and resources, with full cyber-security.

* * * * * * * *

Cyber security for the Smart Grid supports both the reliability of the grid and the confidentiality (and privacy) of the information that is transmitted.

The DOE *Energy Sector-Specific Plan*[9] "envisions a robust, resilient energy infrastructure in which continuity of business and services is maintained through secure and reliable information sharing, effective risk management programs, coordinated response capabilities, and trusted relationships between public and private security partners at all levels of industry and government."

1.2 SCOPE AND DEFINITIONS

The following definition of cyber infrastructure from the National Infrastructure Protection Plan (NIPP) is included to ensure a common understanding.

> **Cyber Infrastructure**: Includes electronic information and communications systems and services and the information contained in these systems and services. Information and communications systems and services are composed of all hardware and software that process, store, and communicate information, or any combination of all of these elements. Processing includes the creation, access, modification, and destruction of information. Storage includes paper, magnetic, electronic, and all other media types. Communications include sharing and distribution of information. For example: computer systems; control systems (e.g., supervisory control and data acquisition–SCADA); networks, such as the Internet; and cyber services (e.g., managed security services) are part of cyber infrastructure.

Traditionally, cyber security for Information Technology (IT) focuses on the protection required to ensure the confidentiality, integrity, and availability of the electronic information communication systems. Cyber security needs to be appropriately applied to the combined power system and IT communication system domains to maintain the reliability of the Smart Grid and privacy of consumer information. Cyber security in the Smart Grid must include a balance of both power and cyber system technologies and processes in IT and power system operations and governance. Poorly applied practices from one domain that are applied into another may degrade reliability.

In the power industry, the focus has been on implementation of equipment that could improve power system reliability. Until recently, communications and IT equipment were typically seen as supporting power system reliability. However, these sectors are becoming more critical to the reliability of the power system. In addition, safety and reliability are of paramount importance in electric power systems. Any cyber security measures in these systems must not impede safe, reliable power system operations.

This report provides guidance to organizations that are addressing cyber security for the Smart Grid (e.g., utilities, regulators, equipment manufacturers and vendors, retail service providers, and electricity and financial market traders). This report is based on what is known at the current time about—

---

[9] Department of Energy, *Energy: Critical Infrastructure and Key Resources, Sector-Specific Plan as input to the National Infrastructure Protection Plan*, May 2007

- The Smart Grid and cyber security;
- Technologies and their use in power systems; and
- Our understanding of the risk environment in which those technologies operate.

This report provides background information on the analysis process used to select and modify the security requirements applicable to the Smart Grid. The process includes both top-down and bottom-up approaches in the selection and modification of security requirements for the Smart Grid. The bottom-up approach focuses on identifying vulnerability classes, for example, buffer overflow and protocol errors. The top-down approach focuses on defining components/domains of the Smart Grid system and the logical interfaces between these components/domains. To reduce the complexity, the logical interfaces are organized into logical interface categories. The inter-component/domain security requirements are specified for these logical interface categories based on the interactions between the components and domains. For example, for the Advanced Metering Infrastructure (AMI) system, some of the security requirements are authentication of the meter to the collector, confidentiality for privacy protection, and integrity for firmware updates.

Finally, this report focuses on Smart Grid operations and not on enterprise operations. However, organizations should capitalize on existing enterprise infrastructures, technologies, support and operational aspects when designing, developing and deploying Smart Grid information systems.

1.3 SMART GRID CYBER SECURITY STRATEGY

The overall cyber security strategy used by the CSWG in the development of this document examined both domain-specific and common requirements when developing a risk mitigation approach to ensure interoperability of solutions across different parts of the infrastructure. The cyber security strategy addressed prevention, detection, response, and recovery. This overall strategy is potentially applicable to other complex infrastructures.

Implementation of a cyber security strategy required the definition and implementation of an overall cyber security risk assessment process for the Smart Grid. *Risk* is the potential for an unwanted outcome resulting from an incident, event, or occurrence, as determined by its likelihood and the associated impacts. This type of risk is one component of organizational risk, which can include many types of risk (e.g., investment risk, budgetary risk, program management risk, legal liability risk, safety risk, inventory risk, and the risk from information systems). The Smart Grid risk assessment process is based on existing risk assessment approaches developed by both the private and public sectors and includes identifying assets, vulnerabilities, and threats and specifying impacts to produce an assessment of risk to the Smart Grid and to its domains and subdomains, such as homes and businesses. Because the Smart Grid includes systems from the IT, telecommunications, and electric sectors, the risk assessment process is applied to all three sectors as they interact in the Smart Grid. The information included in this report is guidance for organizations. NIST is not prescribing particular solutions through the guidance contained in this report. Each organization must develop its own detailed cyber security approach (including a risk assessment methodology) for the Smart Grid.

The following documents were used in developing the risk assessment methodology for the Smart Grid:

- SP 800-39, *DRAFT Managing Risk from Information Systems: An Organizational Perspective*, NIST, April 2008;
- SP 800-30, *Risk Management Guide for Information Technology Systems*, NIST, July 2002;
- Federal Information Processing Standard (FIPS) 200, *Minimum Security Requirements for Federal Information and Information Systems*, NIST, March 2006;
- FIPS 199, *Standards for Security Categorization of Federal Information and Information Systems*, NIST, February 2004;
- *Security Guidelines for the Electricity Sector: Vulnerability and Risk Assessment*, North American Electric Reliability Corporation (NERC), 2002;
- *The National Infrastructure Protection Plan, Partnering to enhance protection and resiliency*, Department of Homeland Security, 2009;
- The IT, telecommunications, and energy sector-specific plans (SSPs), initially published in 2007 and updated annually;
- ANSI/ISA-99.00.01-2007, *Security for Industrial Automation and Control Systems: Concepts, Terminology and Models*, International Society of Automation (ISA), 2007; and
- ANSI/ISA-99.02.01-2009, *Security for Industrial Automation and Control Systems: Establishing an Industrial Automation and Control Systems Security Program*, ISA, January 2009.

The next step in the Smart Grid cyber security strategy was to select and modify (as necessary) the security requirements. The documents used in this step are listed under the description for Task 3. The security requirements and the supporting analyses included in this report may be used by strategists, designers, implementers, and operators of the Smart Grid (e.g., utilities, equipment manufacturers, regulators) as input to their risk assessment process and other tasks in the security lifecycle of the Smart Grid. The information serves as guidance to the various organizations for assessing risk and selecting appropriate security requirements. NIST is not prescribing particular solutions to cyber security issues through the guidance contained in this document.

The cyber security issues that an organization implementing Smart Grid functionality must address are diverse and complicated. This document includes an approach for assessing cyber security issues and selecting and modifying cyber security requirements. Such an approach recognizes that the electric grid is changing from a relatively closed system to a complex, highly interconnected environment, i.e. a system-of-systems. Each organization's implementation of cyber security requirements should evolve as a result of changes in technology and systems, as well as changes in techniques used by adversaries.

The tasks within this cyber security strategy for the Smart Grid were undertaken by participants in the SGIP-CSWG. The remainder of this subsection describes the tasks that have been or will be performed in the implementation of the cyber security strategy. Also included are the deliverables for each task. Because of the time frame within which this report was developed, the tasks listed on the following pages have been performed in parallel, with significant interactions among the groups addressing the tasks.

Figure 1-1 illustrates the tasks defined for the Smart Grid cyber security strategy that are the responsibility of the CSWG. The tasks are defined following the figure.

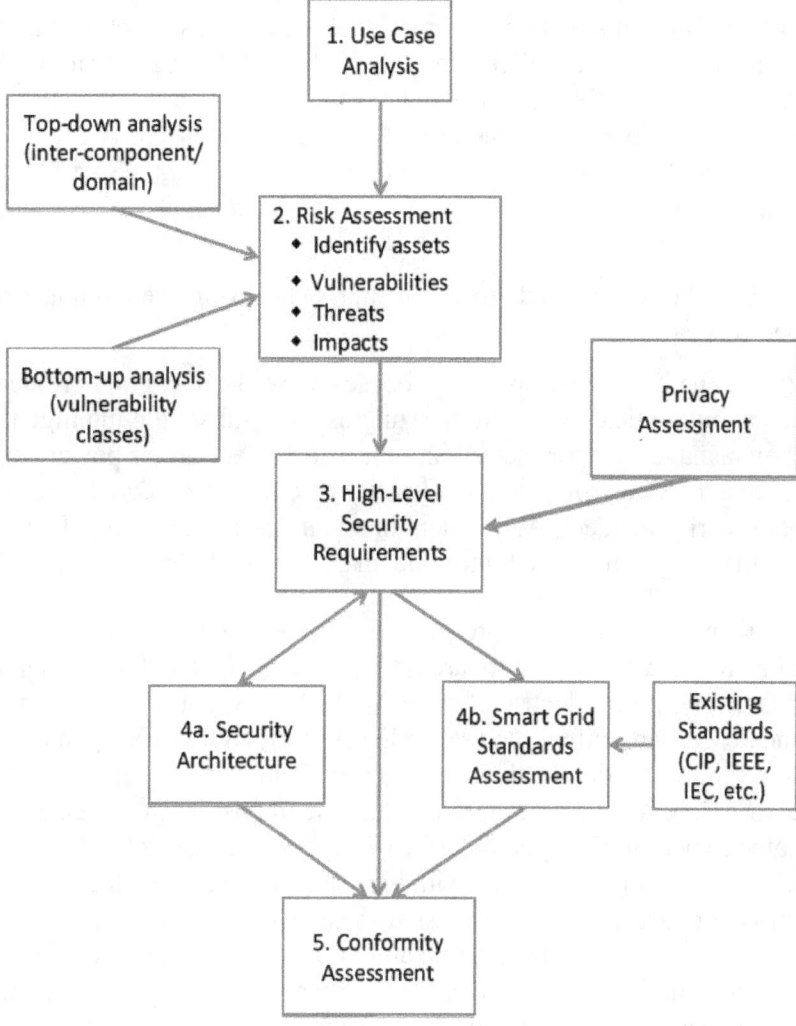

Figure 1-1 Tasks in the Smart Grid Cyber Security Strategy

Task 1. Selection of use cases with cyber security considerations.[10]

The use cases included in Appendix D were selected from several existing sources, e.g., IntelliGrid, Electric Power Research Institute (EPRI) and Southern California Edison (SCE). The set of use cases provides a common framework for performing the risk assessment, developing the logical reference model, and selecting and tailoring the security requirements.

---

[10] A use case is a method of documenting applications and processes for purposes of defining requirements.

Task 2. Performance of a risk assessment

The risk assessment, including identifying assets, vulnerabilities, and threats and specifying impacts has been undertaken from a high-level, overall functional perspective. The output was the basis for the selection of security requirements and the identification of gaps in guidance and standards related to the security requirements.

Vulnerability classes: The initial list of vulnerability classes[11] was developed using information from several existing documents and Web sites, e.g., NIST SP 800-82, Common Weakness Enumeration (CWE) vulnerabilities, and the Open Web Application Security Project (OWASP) vulnerabilities list. These vulnerability classes will ensure that the security controls address the identified vulnerabilities. The vulnerability classes may also be used by Smart Grid implementers, e.g., vendors and utilities, in assessing their systems. The vulnerability classes are included in Chapter 6 of this report.

Overall Analysis: Both bottom-up and top-down approaches were used in implementing the risk assessment as specified earlier.

Bottom-up analysis: The bottom-up approach focuses on well-understood problems that need to be addressed, such as authenticating and authorizing users to substation intelligent electronic devices (IEDs), key management for meters, and intrusion detection for power equipment. Also, interdependencies among Smart Grid domains/systems were considered when evaluating the impacts of a cyber security incident. An incident in one infrastructure can potentially cascade to failures in other domains/systems. The bottom-up analysis is included in Chapter 7 of this report.

Top-down analysis: In the top-down approach, logical interface diagrams were developed for the six functional FERC and NIST priority areas that were the focus of the initial draft of this report—Electric Transportation, Electric Storage, Wide Area Situational Awareness, Demand Response, Advanced Metering Infrastructure, and Distribution Grid Management. This report includes a logical reference model for the overall Smart Grid, with logical interfaces identified for the additional grid functionality. Because there are hundreds of interfaces, each logical interface is allocated to one of 22 logical interface categories. Some examples of the logical interface categories are (1) control systems with high data accuracy and high availability, as well as media and computer constraints; (2) business-to-business (B2B) connections; (3) interfaces between sensor networks and controls systems; and (4) interface to the customer site. A set of attributes (e.g., wireless media, inter-organizational interactions, integrity requirements) was defined and the attributes allocated to the interface categories, as appropriate. This logical interface category/attributes matrix is used in assessing the impact of a security compromise on confidentiality, integrity, and availability. The level of impact is denoted as low, moderate, or high.[12] This assessment was done for each logical interface category. The output from this process was used in the selection of security requirements (Task 3).

As with any assessment, a realistic analysis of the inadvertent errors, acts of nature, and malicious threats and their applicability to subsequent risk-mitigation strategies is critical to the overall outcome. The Smart Grid is no different. It is recommended that all organizations take a

---

[11] A *vulnerability* is a weakness in an information system, system security procedures, internal controls, or implementation that could be exploited or triggered by a threat source. A vulnerability class is a grouping of common vulnerabilities.

[12] The definitions of low, moderate, and high impact are found in FIPS 199.

realistic view of the hazards and threats and work with national authorities as needed to glean the required information, which, it is anticipated, no single utility or other Smart Grid participant would be able to assess on its own. The following table summarizes the categories of adversaries to information systems. These adversaries need to be considered when performing a risk assessment of a Smart Grid information system.

Table 1-1 Categories of Adversaries to Information Systems

Adversary	Description
Nation States	State-run, well organized and financed. Use foreign service agents to gather classified or critical information from countries viewed as hostile or as having an economic, military or a political advantage.
Hackers	A group of individuals (e.g., hackers, phreakers, crackers, trashers, and pirates) who attack networks and systems seeking to exploit the vulnerabilities in operating systems or other flaws.
Terrorists/ Cyberterrorists	Individuals or groups operating domestically or internationally who represent various terrorist or extremist groups that use violence or the threat of violence to incite fear with the intention of coercing or intimidating governments or societies into succumbing to their demands.
Organized Crime	Coordinated criminal activities including gambling, racketeering, narcotics trafficking, and many others. An organized and well-financed criminal organization.
Other Criminal Elements	Another facet of the criminal community, which is normally not well organized or financed. Normally consists of few individuals, or of one individual acting alone.
Industrial Competitors	Foreign and domestic corporations operating in a competitive market and often engaged in the illegal gathering of information from competitors or foreign governments in the form of corporate espionage.
Disgruntled Employees	Angry, dissatisfied individuals with the potential to inflict harm on the Smart Grid network or related systems. This can represent an insider threat depending on the current state of the individual's employment and access to the systems.
Careless or Poorly Trained Employees	Those users who, either through lack of training, lack of concern, or lack of attentiveness pose a threat to Smart Grid systems. This is another example of an insider threat or adversary.

Task 3. Specification of high-level security requirements.

For the assessment of specific security requirements and the selection of appropriate security technologies and methodologies, both cyber security experts and power system experts were needed. The cyber security experts brought a broad awareness of IT and control system security technologies, while the power system experts brought a deep understanding of traditional power system methodologies for maintaining power system reliability.

There are many requirements documents that may be applicable to the Smart Grid. Currently, only NERC Critical Infrastructure Protection (CIP) standards are mandatory for the bulk electric system. The CSWG used three source documents for the cyber security requirements in this report[13]—

---

[13] NIST SP 800-53 is mandatory for federal agencies, and the NERC CIPs are mandatory for the Bulk Power System. This report is a guidance document and is not a mandatory standard.

- NIST SP 800-53, Revision 3, *Recommended Security Controls for Federal Information Systems and Organizations*, August 2009;
- NERC CIP 002, 003-009, version 3; and
- *Catalog of Control Systems Security: Recommendations for Standards Developers*, Department of Homeland Security, March 2010.

These security requirements were then modified for the Smart Grid. To assist in assessing and selecting the requirements, a cross-reference matrix was developed. This matrix, Appendix B, maps the Smart Grid security requirements in this report to the security requirements in SP 800-53, The DHS Catalog, and the NERC CIPs. Each requirement falls in one of three categories: governance, risk and compliance (GRC); common technical; and unique technical. The GRC requirements are applicable to all Smart Grid information systems within an organization and are typically implemented at the organization level and augmented, as required, for specific Smart Grid information systems. The common technical requirements are applicable to all Smart Grid information systems within an organization. The unique technical requirements are allocated to one or more of the logical interface categories defined in the logical reference model included in Chapter 2. Each organization must determine the logical interface categories that are included in each Smart Grid information system. These requirements are provided as guidance and are not mandatory. Each organization will need to perform a risk assessment to determine the applicability of the requirements to their specific situations.

Organizations may find it necessary to identify alternative, but compensating security requirements. A compensating security requirement is implemented by an organization in lieu of a recommended security requirement to provide a comparable level of protection for the information/control system and the information processed, stored, or transmitted by that system. More than one compensating requirement may be required to provide the comparable protection for a particular security requirement. For example, an organization with significant staff limitations may compensate for the recommended separation of duty security requirement by strengthening the audit, accountability, and personnel security requirements within the information/control system. Finally, existing power system capabilities may be used to meet specific security requirements.

Coordination with the Advanced Security Acceleration Project for the Smart Grid: The Advanced Security Acceleration Project for the Smart Grid (ASAP-SG) has made significant contributions to the subgroups that developed this report. ASAP-SG is a utility-driven, public-private collaborative between DOE, the Electric Power Research Institute (EPRI), and a large group of leading North American utilities to develop system-level security requirements for smart grid applications such as advanced metering, third-party access for customer usage data, distribution automation, home area networks, synchrophasors, etc. ASAP-SG is capturing these requirements in a series of Security Profiles, which are submitted to the SG Security Working Group within the UCA International Users Group (UCAIug) for ratification and to the CSWG as input for this report. The collaboration between the CSWG and ASAP-SG has proven most beneficial, as this report provides context and establishes high-level logical interfaces for the ASAP-SG Security Profiles while the Security Profiles provide detailed, actionable, and tailored controls for those building and implementing specific Smart Grid systems.

To date, ASAP-SG has produced two Security Profiles and is nearing completion on a third. The Security Profile for Advanced Metering Infrastructure ("AMI Security Profile") has been ratified by the AMI-SEC Task Force within the UCAIug and provides prescriptive, actionable guidance for how to build-in and implement security from the meter data management system up to and including the home area network interface of the smart meter. The AMI Security Profile served as the basis for early discussions of security for advanced metering functions, eventually informing selection of requirements for the Logical Interface Categories 13 and 14.

The Security Profile for Third Party Data Access ("3PDA Security Profile") is currently under review by a Usability Analysis team within the UCAIug SG Security Working Group, and delineates the security requirements for individuals, utilities, and vendors participating in three-way relationships that involve the ownership and handling of sensitive data (e.g., electric utility customers who want to allow value added service providers to access electric usage data that is in the custody of the customer's utility). The 3PDA Security Profile served as a reference point for many discussions on the subject of privacy, and informed several aspects of Chapter Five – Privacy and the Smart Grid.

Upon completion, the Security Profile for Distribution Management ("DM Security Profile") will address automated distribution management functions including steady state operations and optimization. For this profile "distribution automation" is treated as a specific portion of distribution management related to automated system reconfiguration and SCADA, and is within scope. Publicly available versions of ASAP-SG documentation may be found on SmartGridiPedia at http://www.smartgridipedia.org.

Privacy Impact Assessment: Because the evolving Smart Grid presents potential privacy risks, a privacy impact assessment was performed. Several general privacy principles were used to assess the Smart Grid, and findings and recommendations were developed. The privacy recommendations provide a set of privacy requirements that should be considered when organizations implement Smart Grid information systems. These privacy requirements augment the security requirements specified in Chapter 3.

Task 4a. Development of a logical reference model.

Using the conceptual model included in this report, the FERC and NIST priority area use case diagrams, and the additional areas of AMI and distribution grid management, the CSWG developed a more granular logical reference model for the Smart Grid. This logical reference model consolidates the individual diagrams into a single diagram and expands upon the conceptual model. The additional functionality of the Smart Grid that is not included in the six use case diagrams is included in this logical reference model. The logical reference model identifies logical communication interfaces between actors. This logical reference model is included in Chapter 2 of this report. Because this is a high-level logical reference model, there may be multiple implementations of the logical reference model. In the future, the NIST conceptual model and the logical reference model included in this report will be used by the SGIP Architecture Committee (SGAC) to develop a single Smart Grid architecture. Subsequently, this Smart Grid architecture will be used by the CSWG to revise the logical security architecture included in this report.

Task 4b. Assessment of Smart Grid standards.

In Task 4b, standards that have been identified as potentially relevant to the Smart Grid by the Priority Action Plan (PAP) teams and the SGIP will be assessed to determine relevancy to Smart Grid security. In this process, gaps in security requirements will be identified and recommendations will be made for addressing these gaps. Also, conflicting standards and standards with security requirements not consistent with the security requirements included in this report will be identified with recommendations. This task is ongoing, and the results will be published in a separate document.

Task 5. Conformity Assessment.

The final task is to develop a conformity assessment program for security. This program will be coordinated with the activities defined by the testing and certification standing committee of the SGIP.

1.4 OUTSTANDING ISSUES AND REMAINING TASKS

The following areas need to be addressed in follow-on CSWG activities.

1.4.1 Additional Cyber Security Strategy Areas

Combined cyber-physical attacks: The Smart Grid is vulnerable to coordinated cyber-physical attacks against its infrastructure. Assessing the impact of coordinated cyber-physical attacks will require a sound, risk-based approach because the Smart Grid will inherit all of the physical vulnerabilities of the current power grid (e.g., power outages caused by squirrels). Mitigating physical-only attacks is beyond the scope of this report, which is primarily focused on new risks and vulnerabilities associated with incorporating Smart Grid technologies into the existing power grid. The current version of this document is focused on assessing the impact of cyber-only vulnerabilities.

1.4.2 Future Research and Development (R&D) Topics

There are some R&D themes that are partially addressed in this document that warrant further discussion. There are other R&D themes that are relatively new. The following list consists of topics the R&D group plans to address in the future:

- Synchrophasor Security / NASPInet;
- Anonymization;
- Use of IPv6 in large scale real time control systems;
- Behavioral Economics/Privacy;
- Cross-Domain security involving IT, Power, and Transportation systems; and
- Remote Disablement/Switch of Energy Sources.

1.4.3 Future Cryptography and Key Management Areas

Some topics that will be further developed in the future include:

- Smart Grid adapted PKI: exploration of how to adapt PKI systems for the grid and its various operational and device/system requirements.
- Secure and trusted device profiles: development of a roadmap of different levels of hardware based security functionality that is appropriate for various types of Smart Grid devices.

- Applicable standards: identification and discussion of existing standards that can be used or adapted to meet the cryptography and key management requirements or solve the problems that have been identified.
- Certificate Lifetime: future work should be done to ensure that appropriate guidelines and best practices are established for the Smart Grid community.

1.4.4 Future Privacy Areas

There are privacy concerns for individuals within business premises, such as hotels, hospitals, and office buildings, in addition to privacy concerns for transmitting Smart Grid data across country borders. The privacy use cases included in this report do not address business locations or cross border data transmission. These are topics identified for further investigation.

1.4.5 Roadmap for Vulnerability Classes

The content of the vulnerability chapter is being used across a wide spectrum of industry, from procurement processes in utilities to SDOs and manufacturers, because of the focus on specific and technical analysis that can be responded to with concrete and actionable solutions. This is an encouraging direction for the entire industry. Therefore, we want to encourage the direction of our material becoming more usable across the range of industry. To meet this goal, listed below are some high-level points that will form our roadmap for future activities—

- **Design considerations:** There will be a continued expansion of this material to cover more bottom-up problems and industry issues to provide information that can more directly inform technical elements of procurement processes, as well as specifications and solutions for standards and product development.
- **Specific topics:** Some bottom-up problems and design considerations that began development but were not at a sufficient enough level for inclusion in this version include—
    - Authenticity and trust in the supply chain, and
    - Vulnerability management and traceability in the supply chain.

    The first issue above was driven by the fact that there have been real instances in the broader market with devices that had unauthentic parts or were themselves totally unauthentic. The motives thus far behind these deceptions appeared to be criminal for the sake of economic gain in selling lower cost and quality hardware under the banner of a higher cost and quality brand. This has led to unanticipated failures in the field. This situation brings a strong possibility of reliability issues to the Smart Grid, and if the direction of this threat becomes more malicious with the intent to insert back doors or known flawed components subject to exploitable vulnerability it will elevate the situation to a new level of possible impact.

    Vulnerability management in the supply chain will be focused on the fact that systems and individual devices have become a disparate collection of software and hardware components across very complex supply chains. As a result, it may not be clear to asset owners or the manufacturers directly supplying them the extent to which they may be affected by many reported vulnerabilities in underlying, unknown, and embedded components.

CHAPTER TWO
LOGICAL ARCHITECTURE AND INTERFACES OF THE SMART GRID

This chapter includes a logical reference model of the Smart Grid, including all the major domains—service providers, customer, transmission, distribution, bulk generation, markets, and operations—that are part of the NIST conceptual model. In the future, the NIST conceptual model and the logical reference model included in this report will be used by the SGIP Architecture Committee (SGAC) to develop a single Smart Grid architecture that will be used by the CSWG to revise the logical security architecture included in this report. Figure 2-3 presents the logical reference model and represents a composite high-level view of Smart Grid domains and actors. A Smart Grid domain is a high-level grouping of organizations, buildings, individuals, systems, devices, or other *actors* with similar objectives and relying on—or participating in—similar types of applications.

Communications among actors in the same domain may have similar characteristics and requirements. Domains may contain subdomains. An *actor* is a device, computer system, software program, or the individual or organization that participates in the Smart Grid. Actors have the capability to make decisions and to exchange information with other actors. Organizations may have actors in more than one domain. The actors illustrated in this case are representative examples and do not encompass all the actors in the Smart Grid. Each of the actors may exist in several different varieties and may contain many other actors within them. Table 2-1 complements the logical reference model diagram (Figure 2-3) with a description of the actors associated with the logical reference model.

The logical reference model represents a blending of the initial set of use cases, requirements that were developed at the NIST Smart Grid workshops, the initial NIST Smart Grid Interoperability Roadmap, and the logical interface diagrams for the six FERC and NIST priority areas: electric transportation, electric storage, advanced metering infrastructure (AMI), wide area situational awareness (WASA), distribution grid management, and customer premises.[14] These six priority areas are depicted in individual diagrams with their associated tables. These lower-level diagrams were originally produced at the NIST Smart Grid workshops and then revised for this report. They provide a more granular view of the Smart Grid functional areas. These diagrams are included in Appendix F.

All of the logical interfaces included in the six diagrams are included in the logical reference model. The format for the reference number for each logical interface is UXX, where U stands for universal and XX is the interface number. The reference number is the same on the individual application area diagrams and the logical reference model. This logical reference model focuses on a short-term view (1–3 years) of the proposed Smart Grid and is only a sample representation.

The logical reference model is a work in progress and will be subject to revision and further development. Additional underlying detail as well as additional Smart Grid functions will be needed to enable more detailed analysis of required security functions. The graphic illustrates, at a high level, the diversity of systems as well as a first representation of associations between

---

[14] This was previously named Demand Response.

systems and components of the Smart Grid. The list of actors is a subset of the full list of actors for the Smart Grid and is not intended to be a comprehensive list. This logical reference model is a high-level logical architecture and does not imply any specific implementation.

2.1 THE SEVEN DOMAINS TO THE LOGICAL REFERENCE MODEL

The *NIST Framework and Roadmap* document identifies seven domains within the Smart Grid: Transmission, Distribution, Operations, Bulk Generation, Markets, Customer, and Service Provider. A Smart Grid domain is a high-level grouping of organizations, buildings, individuals, systems, devices, or other *actors* with similar objectives and relying on—or participating in— similar types of applications. The various actors are needed to transmit, store, edit, and process the information needed within the Smart Grid. To enable Smart Grid functionality, the actors in a particular domain often interact with actors in other domains, as shown in Figure 2-1.

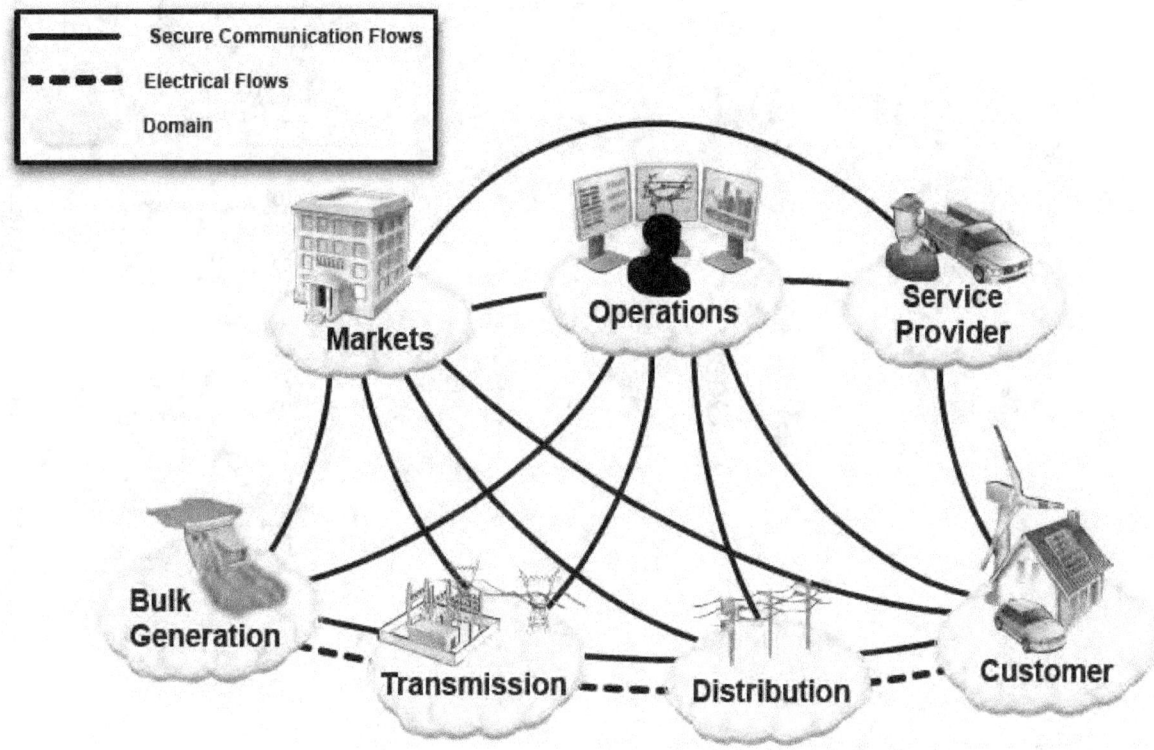

Figure 2-1 Interaction of Actors in Different Smart Grid Domains through Secure Communication Flows

The diagram below (Figure 2-2) expands upon this figure and depicts a composite high-level view of the actors within each of the Smart Grid domains. This high-level diagram is provided as a reference diagram. Actors are devices, systems, or programs that make decisions and exchange information necessary for executing applications within the Smart Grid. The diagrams included later in this chapter expand upon this high-level diagram and include logical interfaces between actors and domains.

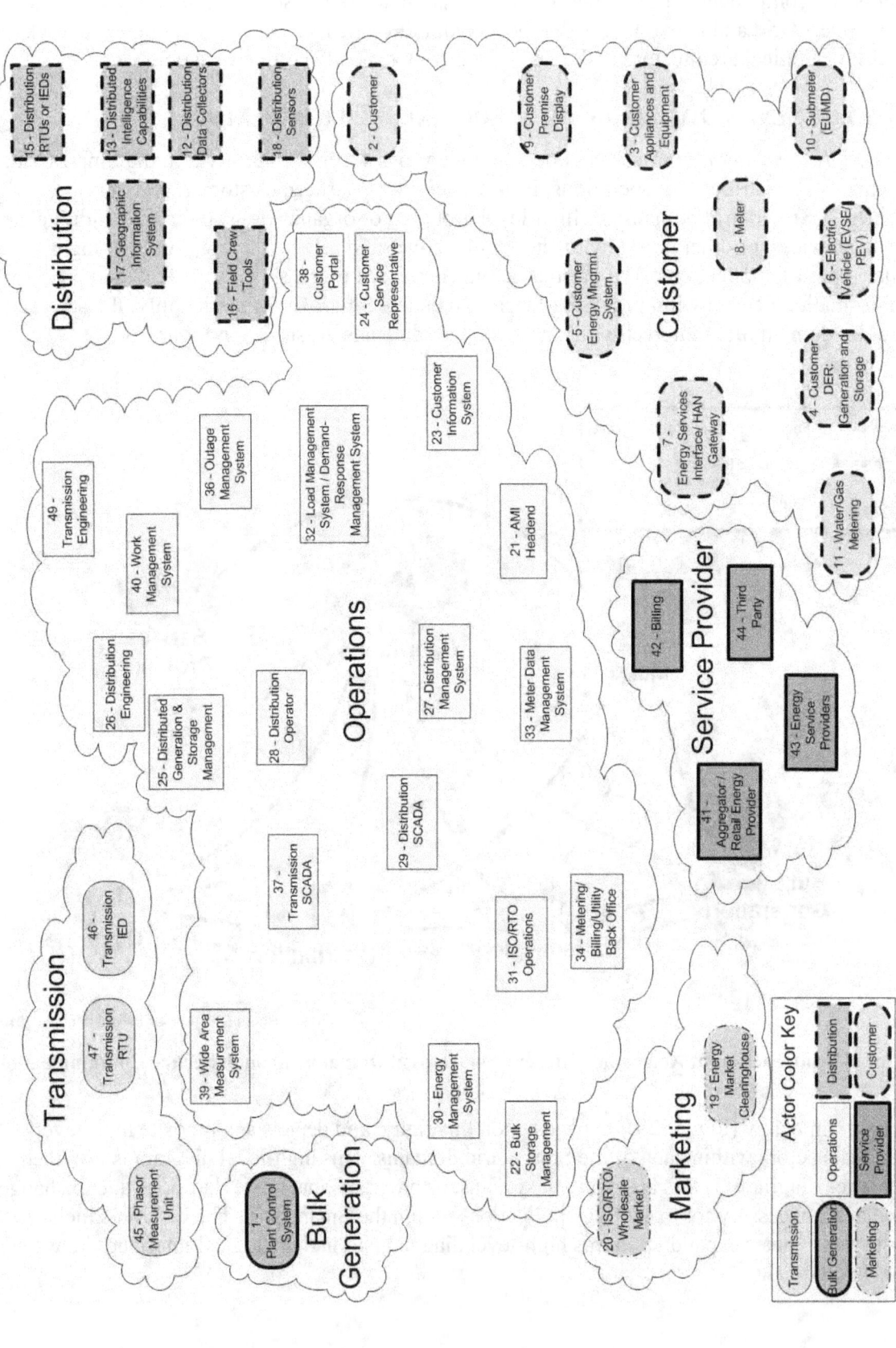

Figure 2-2 Composite High-level View of the Actors within Each of the Smart Grid Domains

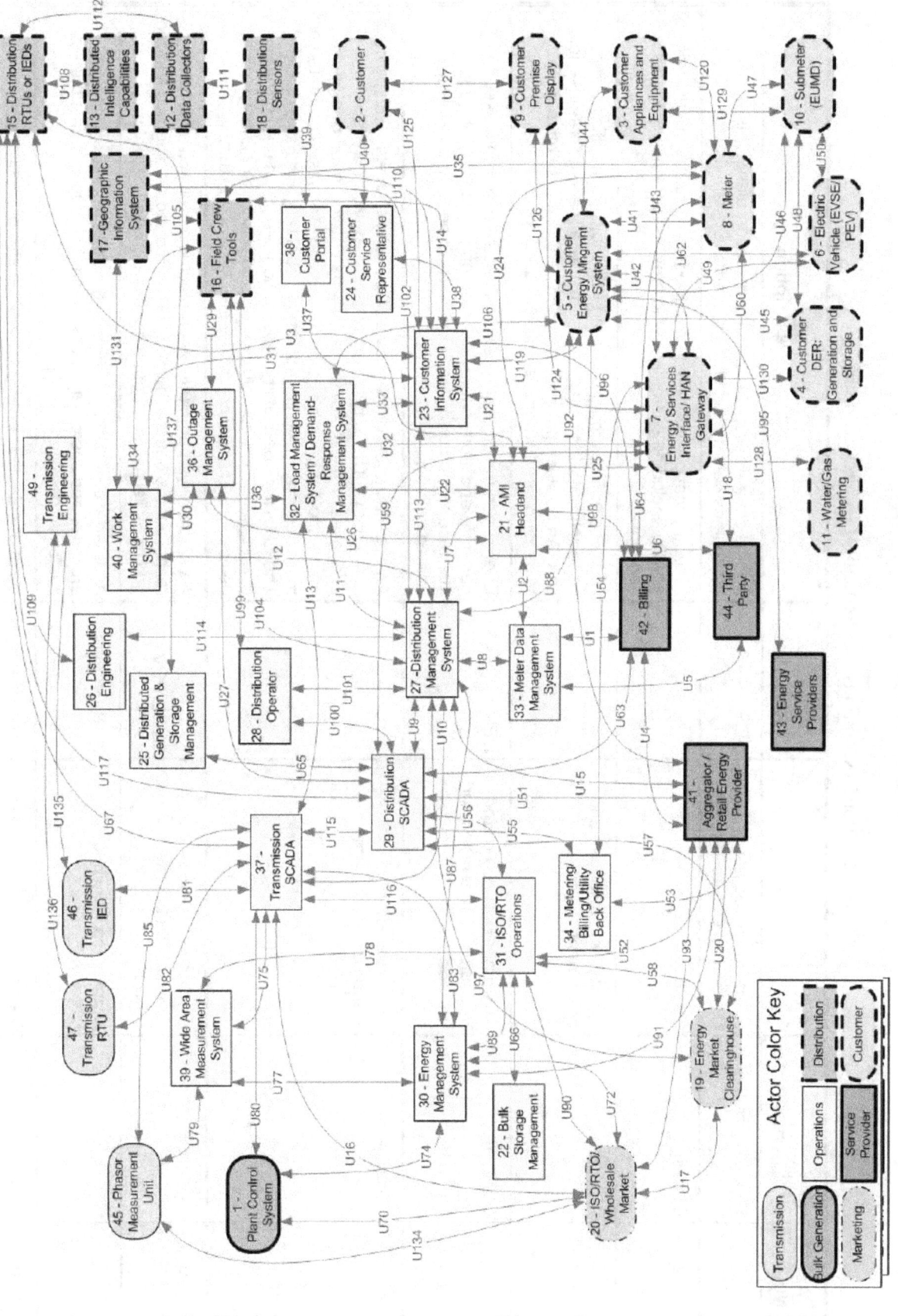

Figure 2-3 Logical Reference Model

Table 2-1 Actor Descriptions for the Logical Reference Model

Actor Number	Domain	Actor	Acronym	Description
1	Bulk Generation	Plant Control System – Distributed Control System	DCS	A local control system at a bulk generation plant. This is sometimes called a Distributed Control System (DCS).
2	Customer	Customer		An entity that pays for electrical goods or services. A customer of a utility, including customers who provide more power than they consume.
3	Customer	Customer Appliances and Equipment		A device or instrument designed to perform a specific function, especially an electrical device, such as a toaster, for household use. An electric appliance or machinery that may have the ability to be monitored, controlled, and/or displayed.
4	Customer	Customer Distributed Energy Resources: Generation and Storage	DER	Energy generation resources, such as solar or wind, used to generate and store energy (located on a customer site) to interface to the controller (HAN/BAN) to perform an energy-related activity.
5	Customer	Customer Energy Management System	EMS	An application service or device that communicates with devices in the home. The application service or device may have interfaces to the meter to read usage data or to the operations domain to get pricing or other information to make automated or manual decisions to control energy consumption more efficiently. The EMS may be a utility subscription service, a third party-offered service, a consumer-specified policy, a consumer-owned device, or a manual control by the utility or consumer.
6	Customer	Electric Vehicle Service Element/Plug-in Electric Vehicle	EVSE/PEV	A vehicle driven primarily by an electric motor powered by a rechargeable battery that may be recharged by plugging into the grid or by recharging from a gasoline-driven alternator.
7	Customer	Home Area Network Gateway	HAN Gateway	An interface between the distribution, operations, service provider, and customer domains and the devices within the customer domain.
8	Customer	Meter		Point of sale device used for the transfer of product and measuring usage from one domain/system to another.

Actor Number	Domain	Actor	Acronym	Description
9	Customer	Customer Premise Display		This device will enable customers to view their usage and cost data within their home or business.
10	Customer	Sub-Meter – Energy Usage Metering Device	EUMD	A meter connected after the main billing meter. It may or may not be a billing meter and is typically used for information-monitoring purposes.
11	Customer	Water/Gas Metering		Point of sale device used for the transfer of product (water and gas) and measuring usage from one domain/system to another.
12	Distribution	Distribution Data Collector		A data concentrator collecting data from multiple sources and modifying/transforming it into different form factors.
13	Distribution	Distributed Intelligence Capabilities		Advanced automated/intelligence application that operates in a normally autonomous mode from the centralized control system to increase reliability and responsiveness.
14	Distribution	Distribution Automation Field Devices		Multifeatured installations meeting a broad range of control, operations, measurements for planning, and system performance reports for the utility personnel.
15	Distribution	Distribution Remote Terminal Unit/Intelligent Electronic Device	RTUs or IEDs	Receive data from sensors and power equipment, and can issue control commands, such as tripping circuit breakers, if they sense voltage, current, or frequency anomalies, or raise/lower voltage levels in order to maintain the desired level.
16	Distribution	Field Crew Tools		A field engineering and maintenance tool set that includes any mobile computing and handheld devices.
17	Distribution	Geographic Information System	GIS	A spatial asset management system that provides utilities with asset information and network connectivity for advanced applications.
18	Distribution	Distribution Sensor		A device that measures a physical quantity and converts it into a signal which can be read by an observer or by an instrument.

Actor Number	Domain	Actor	Acronym	Description
19	Marketing	Energy Market Clearinghouse		Widearea energy market operation system providing high-level market signals for distribution companies (ISO/RTO and Utility Operations). The control is a financial system, not in the sense of SCADA.
20	Marketing	Independent System Operator/Regional Transmission Organization Wholesale Market	ISO/RTO	An ISO/RTO control center that participates in the market and does not operate the market. From the Electric Power Supply Association (EPSA) Web site, "The electric wholesale market is open to anyone who, after securing the necessary approvals, can generate power, connect to the grid and find a counterparty willing to buy their output. These include competitive suppliers and marketers that are affiliated with utilities, independent power producers (IPPs) not affiliated with a utility, as well as some excess generation sold by traditional vertically integrated utilities. All these market participants compete with each other on the wholesale market."[15]
21	Operations	Advanced Metering Infrastructure Headend	AMI	This system manages the information exchanges between third-party systems or systems not considered headend, such as the Meter Data Management System (MDMS) and the AMI network.[16]
22	Operations	Bulk Storage Management		Energy storage connected to the bulk power system.
23	Operations	Customer Information System	CIS	Enterprise-wide software applications that allow companies to manage aspects of their relationship with a customer.
24	Operations	Customer Service Representative	CSR	Customer service provided by a person (e.g., sales and service representative) or by automated means called self-service (e.g., Interactive Voice Response [IVR]).

---

[15] http://www.epsa.org/industry/primer/?fa=wholesaleMarket

[16] Headend (head end)—A central control device required by some networks (e.g., LANs or MANs) to provide such centralized functions as remodulation, retiming, message accountability, contention control, diagnostic control, and access to a gateway. See http://en.wikipedia.org/wiki/Head_end.

Actor Number	Domain	Actor	Acronym	Description
25	Operations	Distributed Generation and Storage Management		Distributed generation is also referred to as on-site generation, dispersed generation, embedded generation, decentralized generation, decentralized energy, or distributed energy. This process generates electricity from many small energy sources for use or storage on dispersed, small devices or systems. This approach reduces the amount of energy lost in transmitting electricity because the electricity is generated very near where it is used, perhaps even in the same building.[17]
26	Operations	Distribution Engineering		A technical function of planning or managing the design or upgrade of the distribution system. For example: • The addition of new customers, • The build out for new load, • The configuration and/or capital investments for improving system reliability.
27	Operations	Distribution Management Systems	DMS	A suite of application software that supports electric system operations. Example applications include topology processor, online three-phase unbalanced distribution power flow, contingency analysis, study mode analysis, switch order management, short-circuit analysis, volt/VAR management, and loss analysis. These applications provide operations staff and engineering personnel additional information and tools to help accomplish their objectives.
28	Operations	Distribution Operator		Person operating the distribution system.
29	Operations	Distribution Supervisory Control and Data Acquisition	SCADA	A type of control system that transmits individual device status, manages energy consumption by controlling compliant devices, and allows operators to directly control power system equipment.

---

[17] Description summarized from http://en.wikipedia.org/wiki/Distributed_generation.

Actor Number	Domain	Actor	Acronym	Description
30	Operations	Energy Management System	EMS	A system of computer-aided tools used by operators of electric utility grids to monitor, controls, and optimize the performance of the generation and/or transmission system. The monitor and control functions are known as SCADA; the optimization packages are often referred to as "advanced applications." (Note: Gas and water could be separate from or integrated within the EMS.)
31	Operations	ISO/RTO Operations		Widearea power system control center providing high-level load management and security analysis for the transmission grid, typically using an EMS with generation applications and network analysis applications.
32	Operations	Load Management Systems/Demand Response Management System	LMS/DRMS	An LMS issues load management commands to appliances or equipment at customer locations in order to decrease load during peak or emergency situations. The DRMS issues pricing or other signals to appliances and equipment at customer locations in order to request customers (or their preprogrammed systems) to decrease or increase their loads in response to the signals.
33	Operations	Meter Data Management System	MDMS	System that stores meter data (e.g., energy usage, energy generation, meter logs, meter test results) and makes data available to authorized systems. This system is a component of the customer communication system. This may also be referred to as a 'billing meter.'

Actor Number	Domain	Actor	Acronym	Description
34	Operations	Metering/Billing/Utility Back Office		Back office utility systems for metering and billing.
36[18]	Operations	Outage Management System	OMS	An OMS is a computer system used by operators of electric distribution systems to assist in outage identification and restoration of power. Major functions usually found in an OMS include: • Listing all customers who have outages. • Prediction of location of fuse or breaker that opened upon failure. • Prioritizing restoration efforts and managing resources based upon criteria such as location of emergency facilities, size of outages, and duration of outages. • Providing information on extent of outages and number of customers impacted to management, media, and regulators. • Estimation of restoration time. • Management of crews assisting in restoration. • Calculation of crews required for restoration.
37	Operations	Transmission SCADA		Transmits individual device status, manages energy consumption by controlling compliant devices, and allowing operators to directly control power system equipment.
38	Operations	Customer Portal		A computer or service that makes available Web pages. Typical services may include: customer viewing of their energy and cost information online, enrollment in prepayment electric services, and enablement of third-party monitoring and control of customer equipment.
39	Operations	Wide Area Measurement System	WAMS	Communication system that monitors all phase measurements and substation equipment over a large geographical base that can use visual modeling and other techniques to provide system information to power system operators.

---

[18] Actor 35 was deleted during development. Actors will be renumbered in the next version of this document.

Actor Number	Domain	Actor	Acronym	Description
40	Operations	Work Management System	WMS	A system that provides project details and schedules for work crews to construct and maintain the power system infrastructure.
41	Service Provider	Aggregator/Retail Energy Provider		Any marketer, broker, public agency, city, county, or special district that combines the loads of multiple end-use customers in facilitating the sale and purchase of electric energy, transmission, and other services on behalf of these customers.
42	Service Provider	Billing		Process of generating an invoice to obtain reimbursement from the customer.
43	Service Provider	Energy Service Provider	ESP	Provides retail electricity, natural gas, and clean energy options, along with energy efficiency products and services.
44	Service Provider	Third Party		A third party providing a business function outside of the utility.
45	Transmission	Phasor Measurement Unit	PMU	Measures the electrical parameters of an electricity grid with respect to universal time (UTC) such as phase angle, amplitude, and frequency to determine the state of the system.
46	Transmission	Transmission IED		IEDs receive data from sensors and power equipment and can issue control commands, such as tripping circuit breakers if they sense voltage, current, or frequency anomalies, or raise/lower voltage levels in order to maintain the desired level. A device that sends data to a data concentrator for potential reformatting.
47	Transmission	Transmission RTU		RTUs pass status and measurement information from a substation or feeder equipment to a SCADA system and transmit control commands from the SCADA system to the field equipment.
48[19]	Operations	Security/Network/System Management		Security/Network/System management devices that monitor and configure the security, network, and system devices.
49	Transmission	Transmission Engineering		Equipment designed for more than 345,000 volts between conductors.

---

[19] Actor 48 is included in logical interface category 22 for security. It is not included in the logical reference model.

2.2 LOGICAL SECURITY ARCHITECTURE OVERVIEW

Smart Grid technologies will introduce millions of new components to the electric grid. Many of these components are critical to interoperability and reliability, will communicate bidirectionally, and will be tasked with maintaining confidentiality, integrity, and availability (CIA) vital to power systems operation.

The definitions of CIA are defined in statue and can be summarized as follows:

Confidentiality: "Preserving authorized restrictions on information access and disclosure, including means for protecting personal privacy and proprietary information...." [44 U.S.C., Sec. 3542]

- A loss of *confidentiality* is the unauthorized disclosure of information.

Integrity: "Guarding against improper information modification or destruction, and includes ensuring information non-repudiation and authenticity...." [44 U.S.C., Sec. 3542]

- A loss of integrity is the unauthorized modification or destruction of information.

Availability: "Ensuring timely and reliable access to and use of information...." [44 U.S.C., SEC. 3542]

- A loss of availability is the disruption of access to or use of information or an information system.

The high-level security requirements address the goals of the Smart Grid. They describe *what* the Smart Grid needs to deliver to enhance security. The logical security architecture describes *where*, at a high level, the Smart Grid will provide security.

This report has identified cyber security requirements for the different logical interface categories. Included in Appendix B are categories of cyber security technologies and services that are applicable to the common technical security requirements. This list of technologies and services is not intended to be prescriptive; rather, it is to be used as guidance.

2.2.1 Logical Security Architecture Key Concepts and Assumptions

A Smart Grid's logical security architecture is constantly in flux because threats and technology evolve. The architecture subgroup specified the following key concepts and assumptions that were the foundation for the logical security architecture.

- **Defense-in-depth strategy:** Security should be applied in layers, with one or more security measures implemented at each layer. The objective is to mitigate the risk of one component of the defense being compromised or circumvented. This is often referred to as "defense-in-depth." A defense-in-depth approach focuses on defending the information (including customer), assets, power systems, and communications and IT infrastructures through layered defenses (e.g., firewalls, intrusion detection systems, antivirus software, and cryptography). Because of the large variety of communication methods and performance characteristics, as well as because no single security measure can counter all types of threats, it is expected that multiple levels of security measures will be implemented.

- **Power system availability:** Power system resiliency to events potentially leading to outages has been the primary focus of power system engineering and operations for decades. Existing power system design and capabilities have been successful in providing this availability for protection against inadvertent actions and natural disasters. These existing power system capabilities may be used to address the cyber security requirements.

The logical security architecture seeks to mitigate threats and threat agents from exploiting system weaknesses and vulnerabilities that can impact the operating environment. A logical security architecture needs to provide protections for data at all interfaces within and among all Smart Grid domains. The logical security architecture baseline assumptions are as follows:

1. A logical security architecture promotes an iterative process for revising the architecture to address new threats, vulnerabilities, and technologies.

2. All Smart Grid systems will be targets.

3. There is a need to balance the impact of a security breach and the resources required to implement mitigating security measures. (Note: The assessment of cost of implementing security is outside the scope of this report. However, this is a critical task for organizations as they develop their cyber security strategy, perform a risk assessment, select security requirements, and assess the effectiveness of those security requirements.)

4. The logical security architecture should be viewed as a business enabler for the Smart Grid to achieve its operational mission (e.g., avoid rendering mission-purposed feature sets inoperative).

5. The logical security architecture is not a one-size-fits-all prescription, but rather a framework of functionality that offers multiple implementation choices for diverse application security requirements within all electric sector organizations.

2.3 LOGICAL INTERFACE CATEGORIES

Each logical interface in the logical reference model was allocated to a logical interface category. This was done because many of the individual logical interfaces are similar in their security-related characteristics and can, therefore, be categorized together as a means to simplify the identification of the appropriate security requirements. These security-related logical interface categories were defined based on attributes that could affect the security requirements.

These logical interface categories and the associated attributes (included in Appendix G) can be used as guidelines by organizations that are developing a cyber security strategy and implementing a risk assessment to select security requirements. This information may also be used by vendors and integrators as they design, develop, implement, and maintain the security requirements. Included below are a listing of all of the logical interfaces by category, the descriptions of each logical interface category, and the associated security architecture diagram. Examples included in the discussions below are not intended to be comprehensive. The user should assess the existing and proposed Smart Grid information system as part of determining which logical interface category should include a specific interface. Listed in each diagram are the unique technical requirements. These security requirements are included in the next chapter.

Table 2-2 Logical Interfaces by Category

Logical Interface Category	Logical Interfaces
1. Interface between control systems and equipment with high availability, and with compute and/or bandwidth constraints, for example: • Between transmission SCADA and substation equipment • Between distribution SCADA and high priority substation and pole-top equipment • Between SCADA and DCS within a power plant	U3, U67, U79, U81, U82, U85, U102, U117, U135, U136, U137
2. Interface between control systems and equipment without high availability, but with compute and/or bandwidth constraints, for example: • Between distribution SCADA and lower priority pole-top equipment • Between pole-top IEDs and other pole-top IEDs	U3, U67, U79, U81, U82, U85, U102, U117, U135, U136, U137
3. Interface between control systems and equipment with high availability, without compute nor bandwidth constraints, for example: • Between transmission SCADA and substation automation systems	U3, U67, U79, U81, U82, U85, U102, U117, U135, U136, U137
4. Interface between control systems and equipment without high availability, without compute nor bandwidth constraints, for example: • Between distribution SCADA and backbone network-connected collector nodes for distribution pole-top IEDs	U3, U67, U79, U81, U82, U85, U102, U117, U135, U136, U137
5. Interface between control systems within the same organization, for example: • Multiple DMS systems belonging to the same utility • Between subsystems within DCS and ancillary control systems within a power plant	U9, U27, U65, U66, U89
6. Interface between control systems in different organizations, for example: • Between an RTO/ISO EMS and a utility energy management system	U7, U10, U13, U16, U56, U74, U80, U83, U87, U115, U116
7. Interface between back office systems under common management authority, for example: • Between a Customer Information System and a Meter Data Management System	U2, U22, U26, U31, U63, U96, U98, U110

Logical Interface Category	Logical Interfaces
8. Interface between back office systems not under common management authority, for example: • Between a third-party billing system and a utility meter data management system	U1, U6, U15, U55
9. Interface with B2B connections between systems usually involving financial or market transactions, for example: • Between a Retail aggregator and an Energy Clearinghouse	U4, U17, U20, U51, U52, U53, U57, U58, U70, U72, U90, U93, U97
10. Interface between control systems and non-control/corporate systems, for example: • Between a Work Management System and a Geographic Information System	U12, U30, U33, U36, U59, U75, U91, U106, U113, U114, U131
11. Interface between sensors and sensor networks for measuring environmental parameters, usually simple sensor devices with possibly analog measurements, for example: • Between a temperature sensor on a transformer and its receiver	U111
12. Interface between sensor networks and control systems, for example: • Between a sensor receiver and the substation master	U108, U112
13. Interface between systems that use the AMI network, for example: • Between MDMS and meters • Between LMS/DRMS and Customer EMS	U8, U21, U25, U32, U95, U119, U130
14. Interface between systems that use the AMI network with high availability, for example: • Between MDMS and meters • Between LMS/DRMS and Customer EMS • Between DMS Applications and Customer DER • Between DMS Applications and DA Field Equipment	U8, U21, U25, U32, U95, U119, U130
15. Interface between systems that use customer (residential, commercial, and industrial) site networks which include: • Between Customer EMS and Customer Appliances • Between Customer EMS and Customer DER • Between Energy Service Interface and PEV	U42, U43, U44, U45, U49, U62, U120, U124, U126, U127

Logical Interface Category	Logical Interfaces
16. Interface between external systems and the customer site, for example: • Between Third Party and HAN Gateway • Between ESP and DER • Between Customer and CIS Web site	U18, U37, U38, U39, U40, U88, U92, U100, U101, U125
17. Interface between systems and mobile field crew laptops/equipment, for example: • Between field crews and GIS • Between field crews and substation equipment	U14, U29, U34, U35, U99, U104, U105
18. Interface between metering equipment, for example: • Between sub-meter to meter • Between PEV meter and Energy Service Provider	U24, U41, U46, U47, U48, U50, U54, U60, U64, U128, U129
19. Interface between operations decision support systems, for example: • Between WAMS and ISO/RTO	U77, U78, U134
20. Interface between engineering/maintenance systems and control equipment, for example: • Between engineering and substation relaying equipment for relay settings • Between engineering and pole-top equipment for maintenance • Within power plants	U11, U109
21. Interface between control systems and their vendors for standard maintenance and service, for example: • Between SCADA system and its vendor	U5
22. Interface between security/network/system management consoles and all networks and systems, for example: • Between a security console and network routers, firewalls, computer systems, and network nodes	U133 (includes interfaces to actors 17- Geographic Information System, 12 – Distribution Data Collector, 38 – Customer Portal, 24 – Customer Service Representative, 23 – Customer Information System, 21 – AMI Headend, 42 – Billing, 44 – Third Party, 43 – Energy Service Provider, 41 – Aggregator / Retail Energy Provider, 19 – Energy Market Clearinghouse, 34 – Metering / Billing / Utility Back Office)

2.3.1 Logical Interface Categories 1—4

Logical Interface Category 1: Interface between control systems and equipment with high availability, and with compute and/or bandwidth constraints

Logical Interface Category 2: Interface between control systems and equipment without high availability, but with compute and/or bandwidth constraints

Logical Interface Category 3: Interface between control systems and equipment with high availability, without compute or bandwidth constraints

Logical Interface Category 4: Interface between control systems and equipment without high availability, without compute or bandwidth constraints

Logical interface categories 1 through 4 cover communications between control systems (typically centralized applications such as a SCADA master station) and equipment as well as communications between equipment. The equipment is categorized with or without high availability. The interface communication channel is categorized with or without computational and/or bandwidth constraints. All activities involved with logical interface categories 1 through 4 are typically machine-to-machine actions. Furthermore, communication modes and types are similar between logical interface categories 1 through 4 and are defined as follows:

- Interface Data Communication Mode
  - Near Real-Time Frequency Monitoring Mode (ms, subcycle based on a 60 Hz system) (may or may not include control action communication)
  - High Frequency Monitoring Mode (2 s ≤ 60 s scan rates)
  - Low Frequency Monitoring Mode (scan/update rates in excess of 1 min, file transfers)
- Interface Data Communication Type
  - Monitoring and Control Data for real-time control system environment (typical measurement and control points)
  - Equipment Maintenance and Analysis (numerous measurements on field equipment that is typically used for preventive maintenance and post analysis)
  - Equipment Management Channel (remote maintenance of equipment)

The characteristics that vary between and distinguish each logical interface category are the availability requirements for the interface and the computational/communications constraints for the interface as follows:

- Availability Requirements – Availability requirements will vary between these interfaces and are driven primarily by the power system application which the interface supports and not by the interface itself. For example, a SCADA interface to a substation or pole-top RTU may have a HIGH availability requirement in one case because it is supporting critical monitoring and switching functions or a MODERATE to LOW availability if supporting an asset-monitoring application.

- Communications and Computational Constraints — Computational constraints are associated with cryptography requirements on the interface. The use of cryptography typically has high CPU needs for mathematical calculations, although it is feasible to implement cryptographic processing in peripheral hardware. Existing devices like RTUs, substation IEDs, meters, and others are typically not equipped with sufficient digital hardware to perform cryptography or other security functions.
- Bandwidth constraints are associated with data volume on the interface. In this case, media is usually narrowband, limiting the volume of traffic, and impacting the types of security measures that are feasible.

With these requirements and constraints, logical interface categories 1 through 4 can be defined as follows:

1. Interface between control systems and equipment with high availability and with computational and/or bandwidth constraints:
   - Between transmission SCADA in support of state estimation and substation equipment for monitoring and control data using a high frequency mode;
   - Between distribution SCADA in support of three phase, real-time power flow and substation equipment for monitoring data using a high and low frequency mode;
   - Between transmission SCADA in support of automatic generation control (AGC) and DCS within a power plant for monitoring and control data using a high frequency mode;
   - Between SCADA in support of Volt/VAR control and substation equipment for monitoring and control data using a high and low frequency mode; and
   - Between transmission SCADA in support of contingency analysis and substation equipment for monitoring data using high frequency mode.

2. Interface between control systems and equipment without high availability and with computational and/or bandwidth constraints:
   - Between field devices and control systems for analyzing power system faults using a low frequency mode;
   - Between a control system historian and field devices for capturing power equipment attributes using a high or low frequency mode;
   - Between distribution SCADA and lower priority pole-top devices for monitoring field devices using a low frequency mode; and
   - Between pole-top IEDs and other pole-top IEDs (not used of protection or automated switching) for monitoring and control in a high or low frequency mode.

3. Interface between control systems and equipment with high availability without computational and/or bandwidth constraints:
   - Between transmission SCADA and substation automation systems for monitoring and control data using a high frequency mode;

- Between EMS and generation control (DCS) and RTUs for monitoring and control data using a high frequency mode;
- Between distribution SCADA and substation automation systems, substation RTUs, and pole-top devices for monitoring and control data using a high frequency mode;
- Between a PMU device and a phasor data concentrator (PDC) for monitoring data using a high frequency mode; and
- Between IEDs (peer-to-peer) for power system protection, including transfer trip signals between equipment in different substations.

4. Interface between control systems and equipment without high availability, without computational and/or bandwidth constraints:
   - Between field device and asset monitoring system for monitoring data using a low frequency mode;
   - Between field devices (relays, digital fault recorders [DFRs], power quality [PQ]) and event analysis systems for event, disturbance, and PQ data;
   - Between distribution SCADA and lower-priority pole-top equipment for monitoring and control data in a high or low frequency mode;
   - Between pole-top IEDs and other pole-top IEDs (not used for protection or automated switching) for monitoring and control in a high or low frequency mode; and
   - Between distribution SCADA and backbone network-connected collector nodes for lower-priority distribution pole-top IEDs for monitoring and control in a high or low frequency mode.

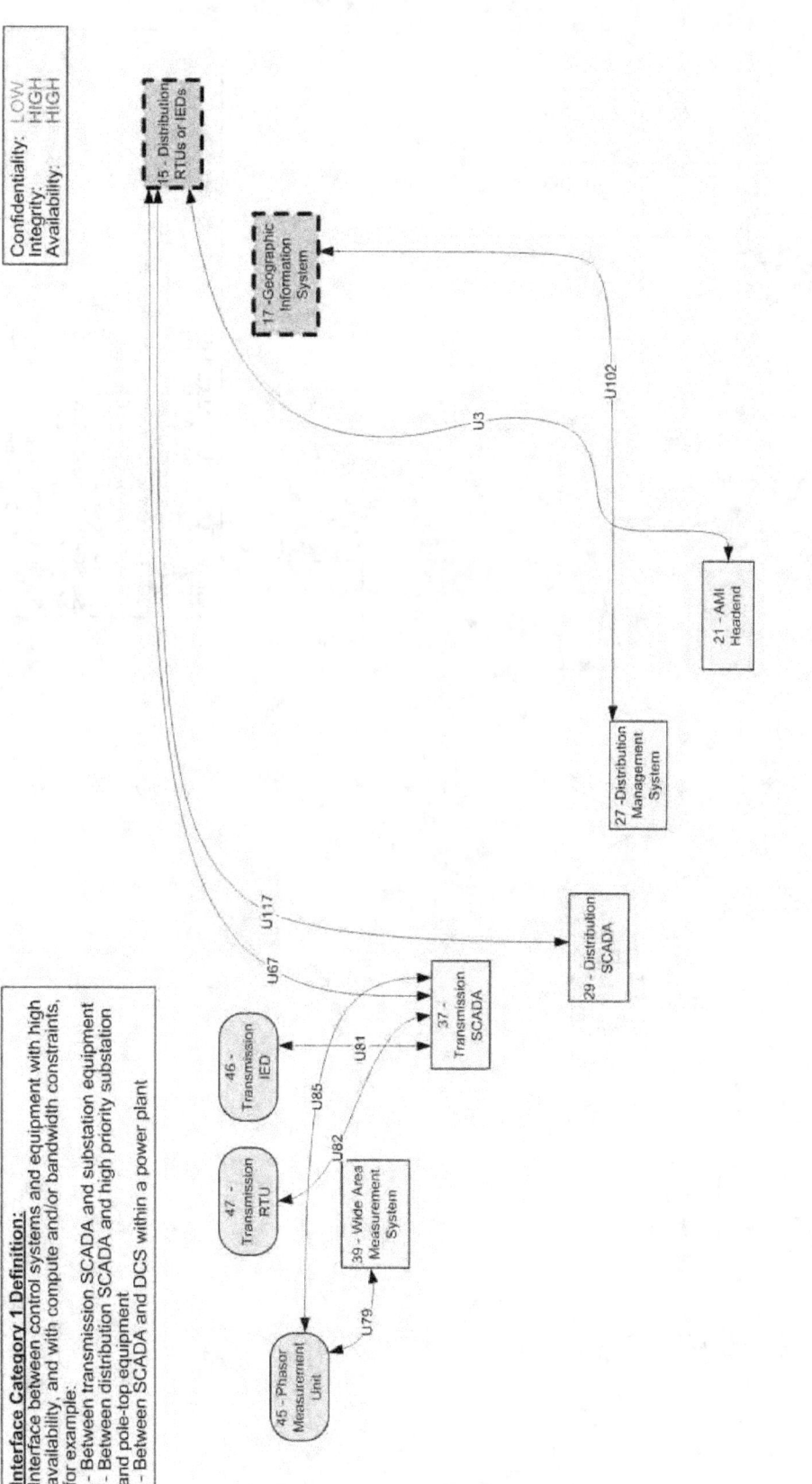

Figure 2-4 Logical Interface Category 1

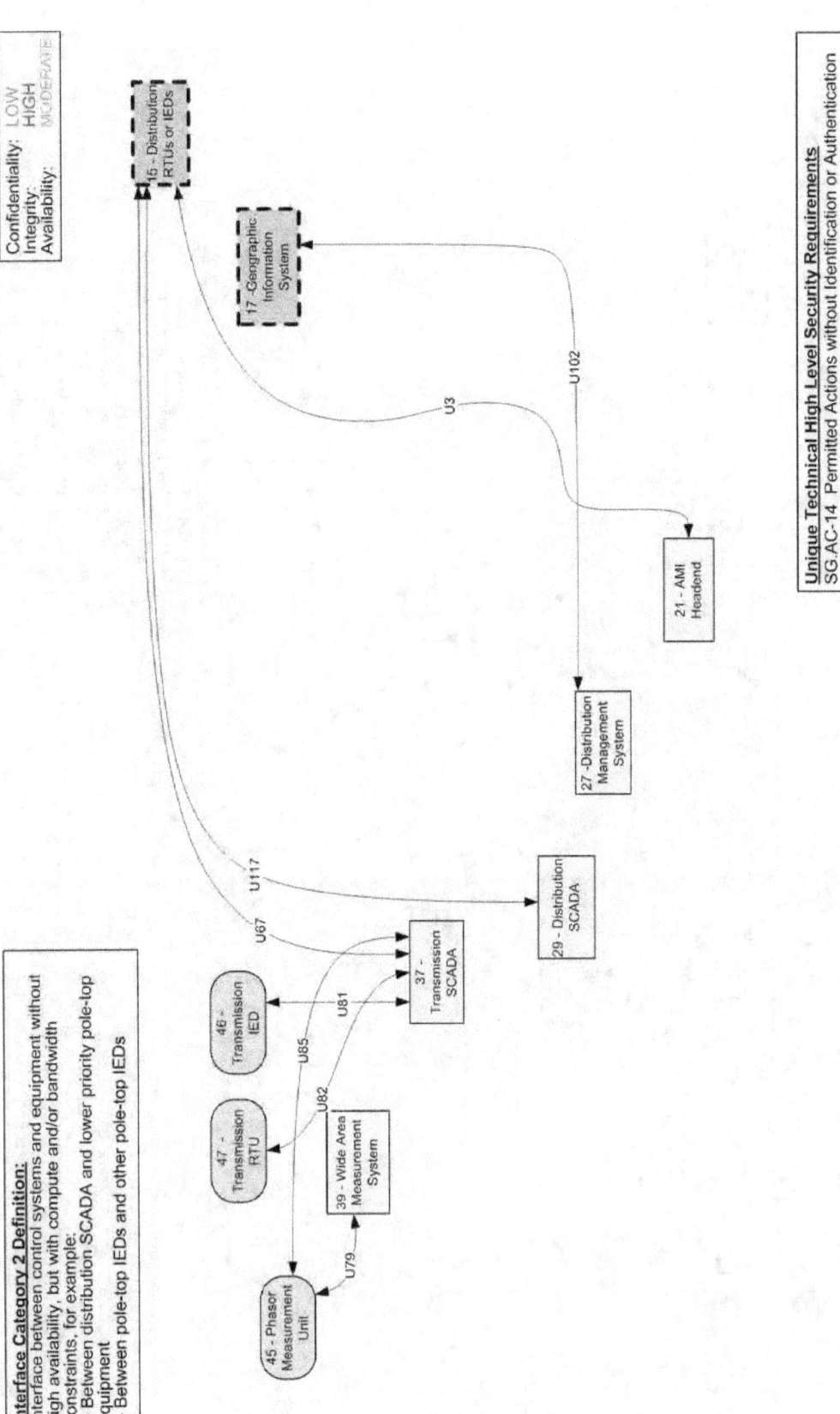

Figure 2-5 Logical Interface Category 2

NISTIR 7628 Guidelines for Smart Grid Cyber Security v1.0 – Aug 2010

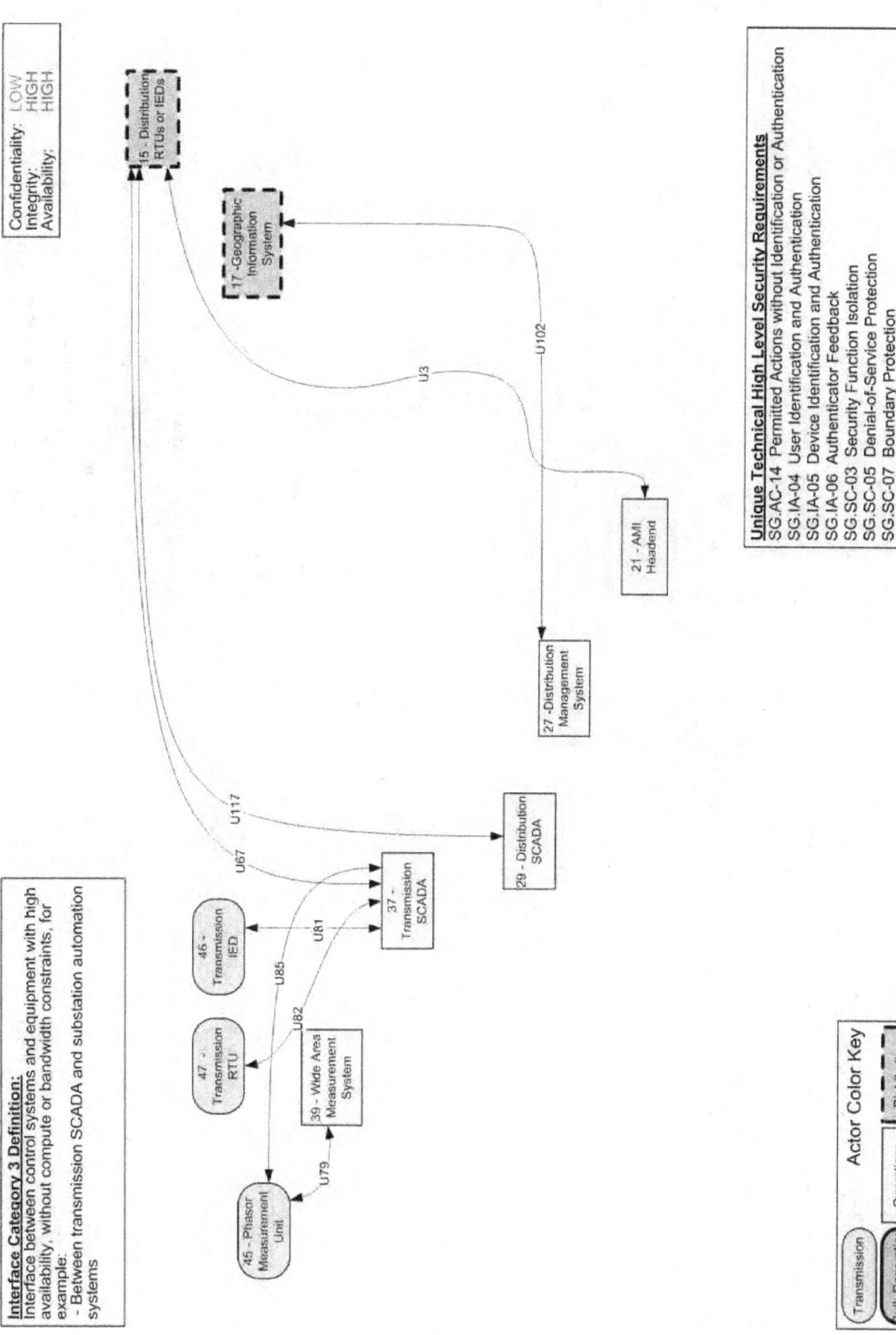

Figure 2-6 Logical Interface Category 3

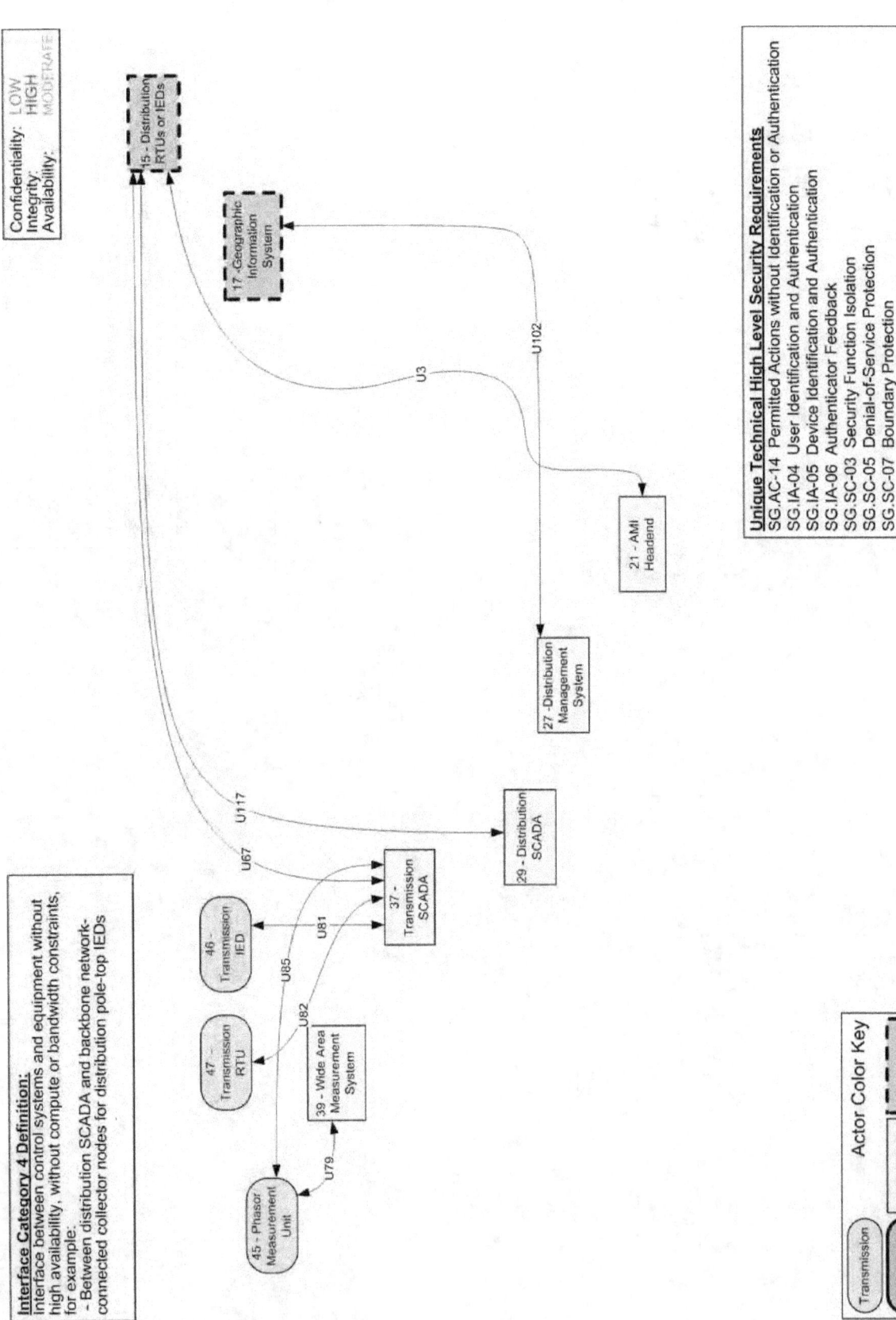

Figure 2-7 Logical Interface Category 4

2.3.2 Logical Interface Category 5: Interface between control systems within the same organization

Logical interface category 5 covers the interfaces between control systems within the same organization, for example:

- Between multiple data management systems belonging to the same utility; and
- Between subsystems within DCS and ancillary control systems within a power plant.

Control systems with interfaces between them have the following characteristics and issues:

- Since control systems generally have high data accuracy and high availability requirements, the interfaces between them need to implement those security requirements even if they do not have the same requirements.
- The interfaces generally use communication channels (wide area networks [WANs] and/or local area networks [LANs]) that are designed for control systems.
- The control systems themselves are usually in secure environments, such as within a utility control center or within a power plant.

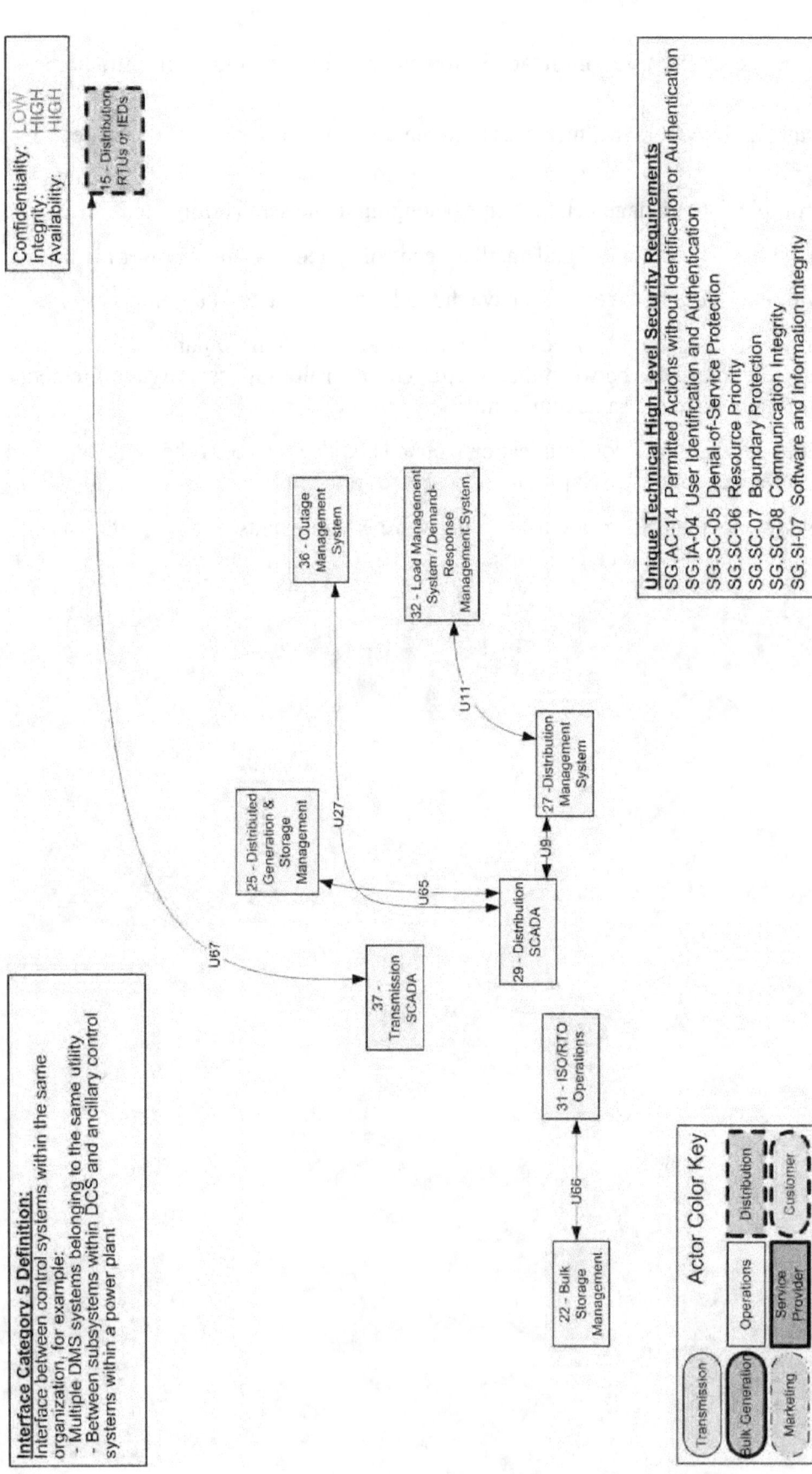

Figure 2-8 Logical Interface Category 5

2.3.3 Logical Interface Category 6: Interface between control systems in different organizations

Logical interface category 6 covers the interfaces between control systems in different organizations, for example:

- Between an RTO/ISO EMS and a utility energy management system;
- Between a Generation and Transmission (G&T) SCADA and a distribution CO-OP SCADA;
- Between a transmission EMS and a distribution DMS in different utilities; and
- Between an EMS/SCADA and a power plant DCS.

Control systems with interfaces between them have the following characteristics and issues:

- Since control systems generally have high data accuracy and high availability requirements, the interfaces between them need to implement those security requirements even if they do not have the same requirements.
- The interfaces generally use communication channels (WANs and/or LANs) that are designed for control systems.
- The control systems are usually in secure environments, such as within a utility control center or within a power plant.
- Since the control systems are in different organizations, the establishment and maintenance of the chain of trust is more important.

NISTIR 7628 Guidelines for Smart Grid Cyber Security v1.0 – Aug 2010

Interface Category 6 Definition:
Interface between control systems in different organizations, for example:
- Between an RTO/ISO EMS and a utility energy management system

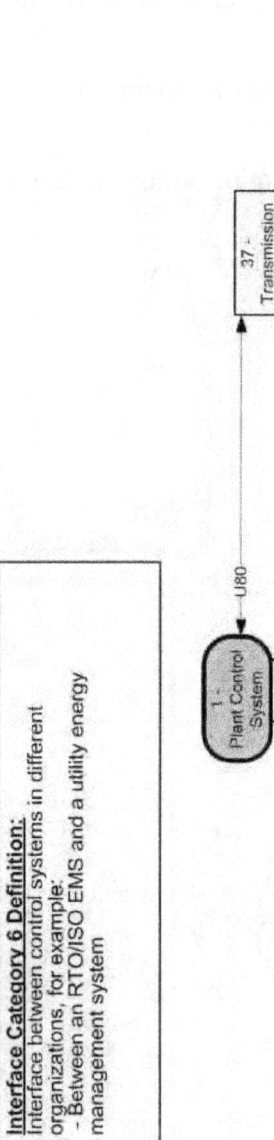
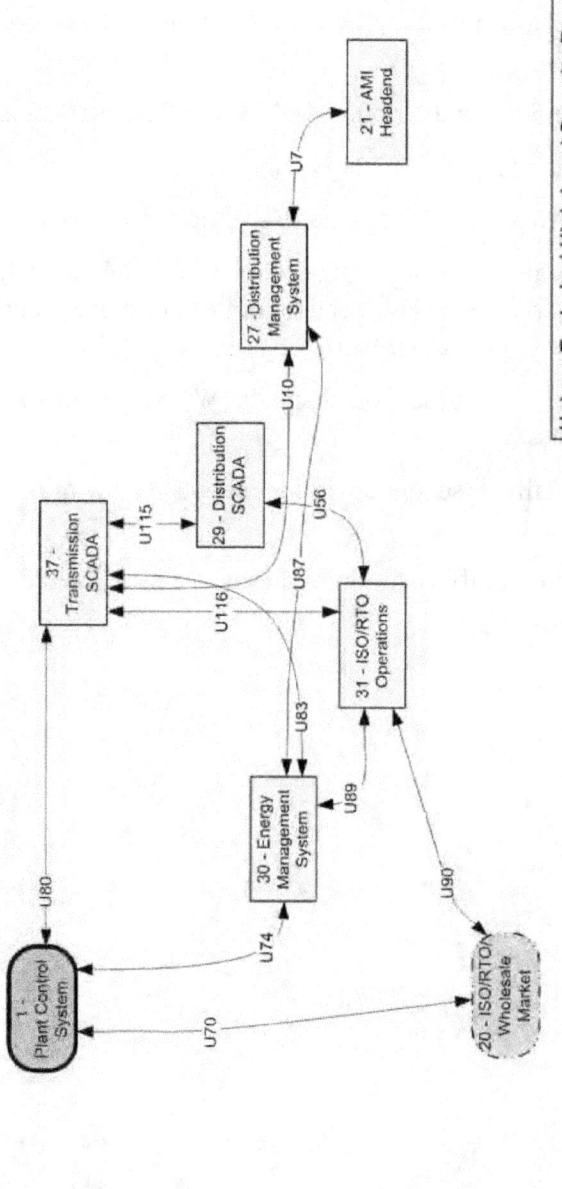
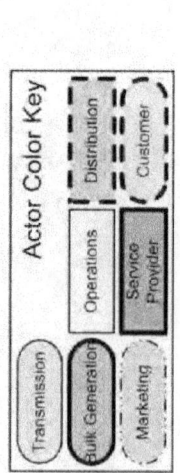

Confidentiality: LOW
Integrity: HIGH
Availability: MODERATE

Unique Technical High Level Security Requirements
SG.AC-14 Permitted Actions without Identification or Authentication
SG.IA-04   User Identification and Authentication
SG.SC-05 Denial-of-Service Protection
SG.SC-06 Resource Priority
SG.SC-07 Boundary Protection
SG.SC-08 Communication Integrity
SG.SI-07   Software and Information Integrity

Figure 2-9 Logical Interface Category 6

40

2.3.4 Logical Interface Categories 7—8

Logical Interface Category 7: Interface between back office systems under common management authority

Logical Interface Category 8: Interface between back office systems not under common management authority

Logical interface category 7 covers the interfaces between back office systems that are under common management authority, e.g., between a CIS and a MDMS. Logical interface category 8 covers the interfaces between back office systems that are not under common management authority, e.g., between a third-party billing system and a utility MDMS. These logical interface categories are focused on confidentiality and privacy rather than on power system reliability.

NISTIR 7628 Guidelines for Smart Grid Cyber Security v1.0 – Aug 2010

Interface Category 7 Definition:
Interface between back office systems under common management authority, for example:
- Between a Customer Information System and a Meter Data Management System

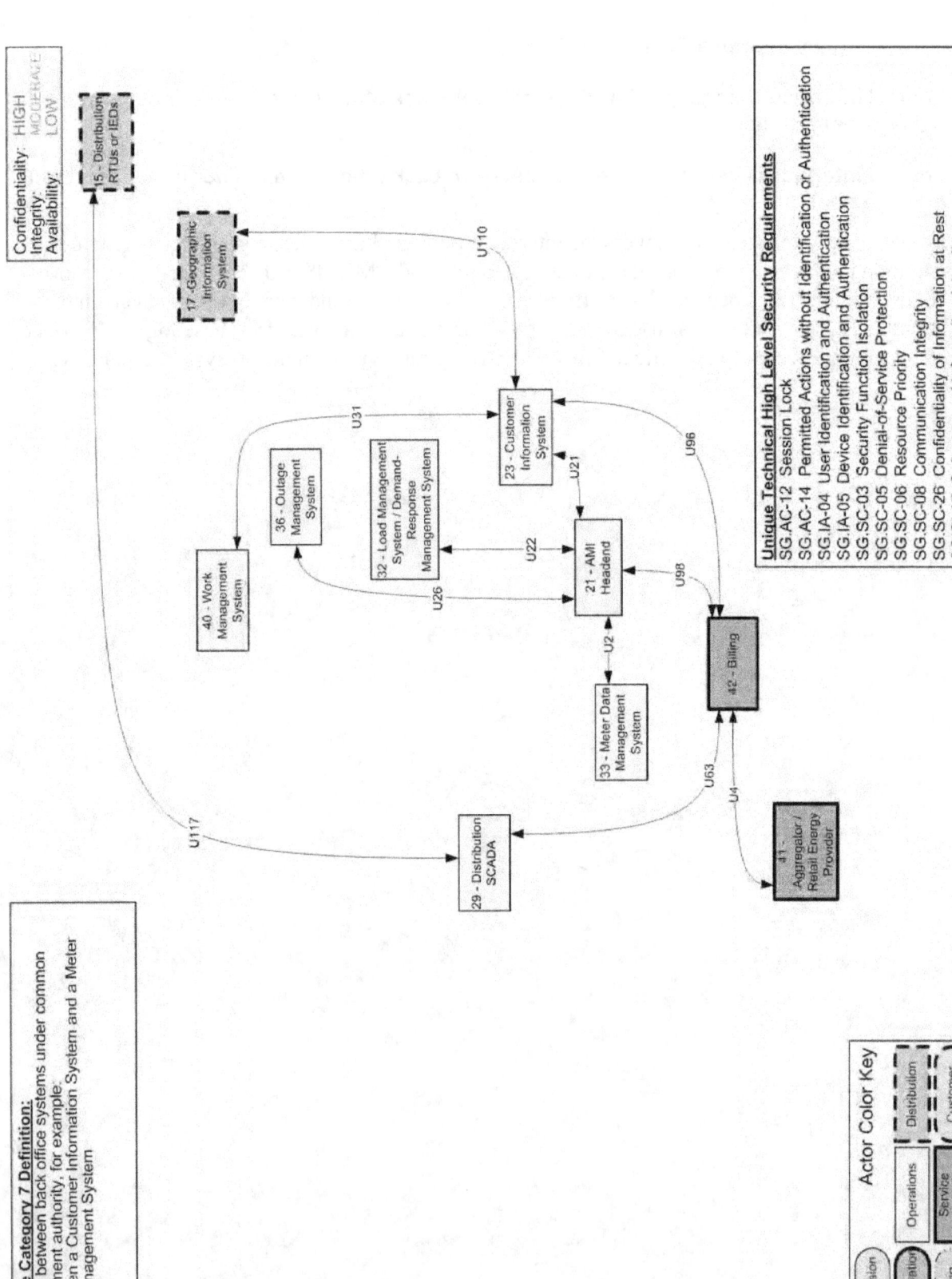

Figure 2-10 Logical Interface Category 7

42

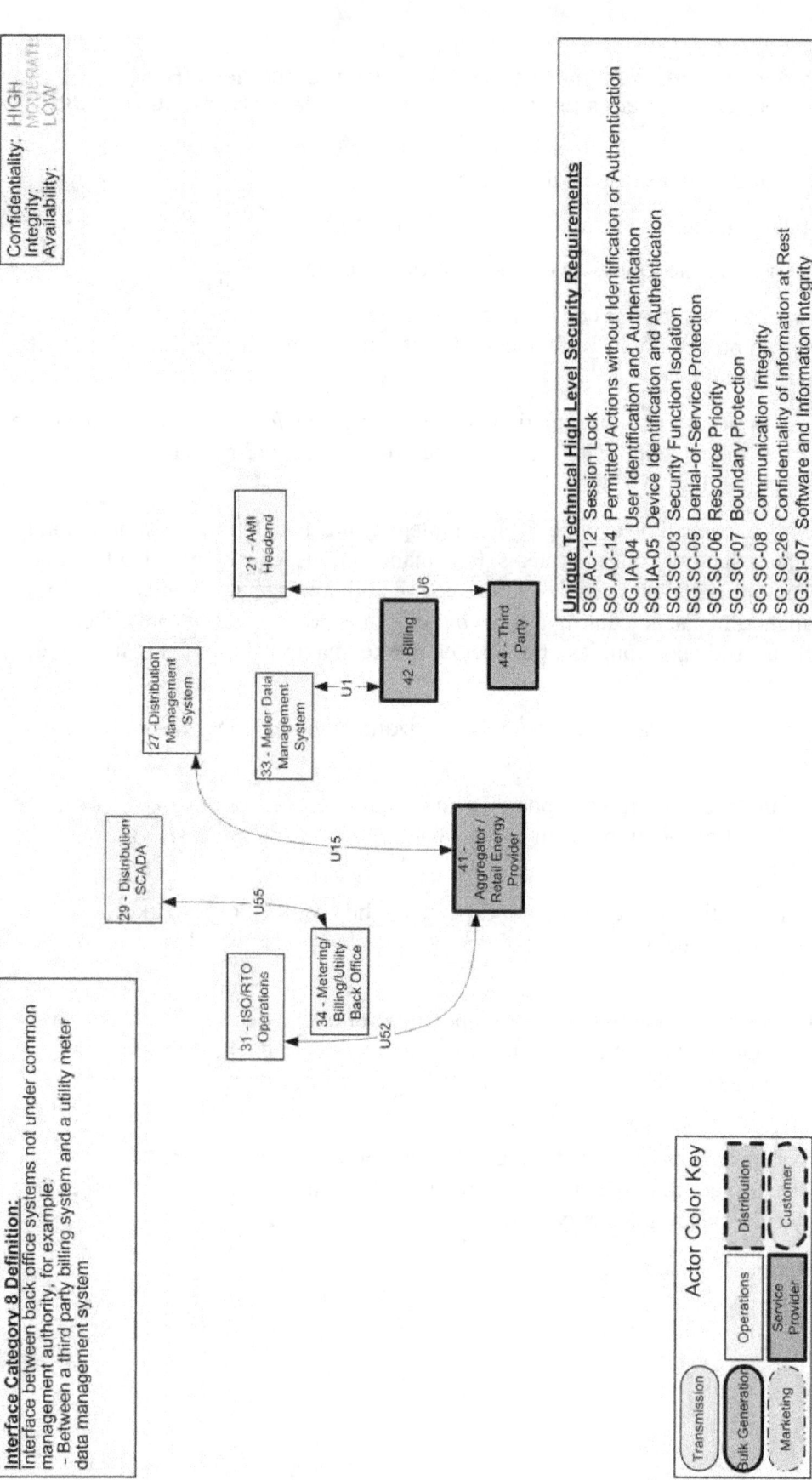

Figure 2-11 Logical Interface Category 8

2.3.5 Logical Interface Category 9: Interface with business to business (B2B) connections between systems usually involving financial or market transactions

Logical interface category 9 covers the interface with B2B connections between systems usually involving financial or market transactions, for example:

- Between a retail aggregator and an energy clearinghouse.

These B2B interactions have the following characteristics and issues:

- Confidentiality needs to be considered since the interactions involve financial transactions with potentially large financial impacts and where confidential bids are vital to a legally operating market.

- Privacy, in terms of historical information on what energy and/or ancillary services were bid, is important to maintaining legal market operations and avoiding market manipulation or gaming.

- Timing latency (critical time availability) and integrity are also important, although in a different manner than for control systems. For financial transactions involving bidding into a market, timing can be crucial. Therefore, although average availability does not need to be high, time latency during critical bidding times is crucial to avoid either inadvertently missed opportunities or deliberate market manipulation or gaming of the system.

- By definition, market operations are across organizational boundaries, thus posing trust issues.

- It is expected that many customers, possibly through aggregators or other energy service providers, will participate in the retail energy market, thus vastly increasing the number of participants.

- Special communication networks are not expected to be needed for the market transactions and may include the public Internet as well as other available wide area networks.

- Although the energy market has now been operating for over a decade at the bulk power level, the retail energy market is in its infancy. Its growth over the next few years is expected, but no one yet knows in what directions or to what extent that growth will occur.

- However, systems and procedures for market interactions are very mature industry concepts. The primary requirement, therefore, is to utilize those concepts and protections in the newly emerging retail energy market.

NISTIR 7628 Guidelines for Smart Grid Cyber Security v1.0 – Aug 2010

Interface Category 9 Definition:
Interface with B2B connections between systems usually involving financial or market transactions, for example:
- Between a Retail aggregator and an Energy Clearinghouse

Confidentiality: LOW
Integrity: MODERATE
Availability: MODERATE

Unique Technical High Level Security Requirements
SG.AC-14 Permitted Actions without Identification or Authentication
SG.IA-04 User Identification and Authentication
SG.SC-05 Denial-of-Service Protection
SG.SC-06 Resource Priority
SG.SC-07 Boundary Protection
SG.SC-08 Communication Integrity
SG.SC-09 Communication Confidentiality
SG.SI-07 Software and Information Integrity

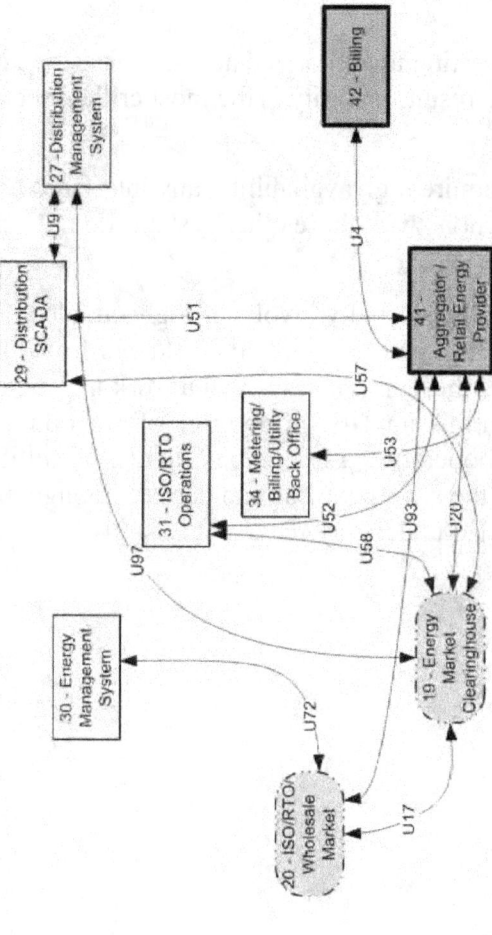

Figure 2-12 Logical Interface Category 9

2.3.6 Logical Interface Category 10: Interface between control systems and non-control/corporate systems

Logical interface category 10 covers the interfaces between control systems and non-control/corporate systems, for example:

- Between a WMS and a GIS;
- Between a DMS and a CIS;
- Between an OMS and the AMI headend system; and
- Between an OMS and a WMS.

These interactions between control systems and non-control systems have the following characteristics and issues:

- The primary security issue is preventing unauthorized access to sensitive control systems through non-control systems. As a result, integrity is the most critical security requirement.

- Since control systems generally require high availability, any interfaces with non-control systems should ensure that interactions with these other systems do not compromise the high reliability requirement.

- The interactions between these systems usually involve loosely coupled interactions with very different types of exchanges from one system to the next and from one vendor to the next. Therefore, standardization of these interfaces is still a work in progress, with the International Electrotechnical Commission (IEC) Common Information Model (CIM)[20] and the National Rural Electric Cooperative Association (NRECA) MultiSpeak® specification expected to become the most common standards, although other efforts for special interfaces (e.g., GIS) are also under way.

---

[20] IEC 61970/69 Common Information Model.

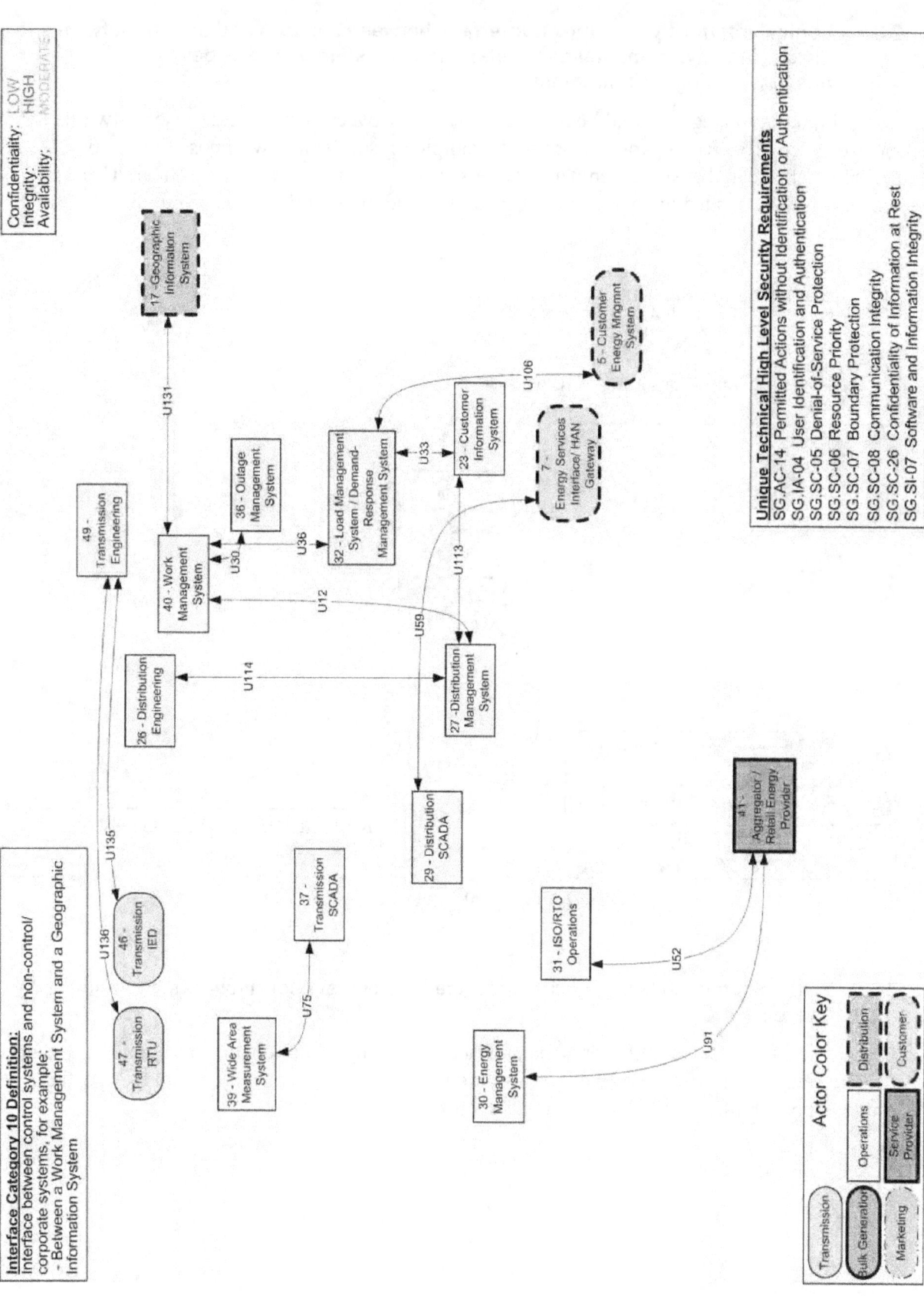

Figure 2-13 Logical Interface Category 10

2.3.7 Logical Interface Category 11: Interface between sensors and sensor networks for measuring environmental parameters, usually simple sensor devices with possibly analog measurements

Logical interface category 11 addresses the interfaces between sensors and sensor networks for measuring environmental parameters, usually simple sensor devices with possibly analog measurements, e.g., between a temperature sensor on a transformer and its receiver. These sensors are very limited in computational capability and often limited in communication bandwidth.

Interface Category 11 Definition:
Interface between sensors and sensor networks for measuring environmental parameters, usually simple sensor devices with possibly analog measurements, for example:
- Between a temperature sensor on a transformer and its receiver

Confidentiality: LOW
Integrity: MODERATE
Availability: MODERATE

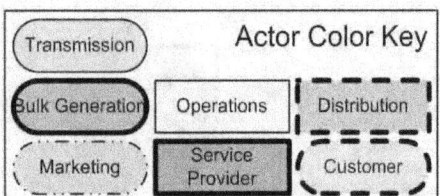

Unique Technical High Level Security Requirements
SG.SC-08 Communication Integrity

Figure 2-14 Logical Interface Category 11

2.3.8 Logical Interface Category 12: Interface between sensor networks and control systems

Logical interface category 12 addresses interfaces between sensor networks and control systems, e.g., between a sensor receiver and the substation master. These sensor receivers are usually limited in capabilities other than collecting sensor information.

Interface Category 12 Definition:
Interface between sensor networks and control systems, for example:
- Between a sensor receiver and the substation master

Confidentiality: LOW
Integrity: MODERATE
Availability: MODERATE

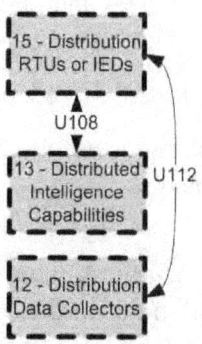

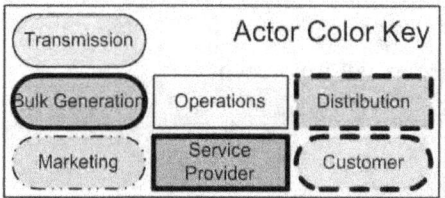

Unique Technical High Level Security Requirements
SG.IA-06 Authenticator Feedback
SG.IA-05 Device Identification and Authentication
SG.SC-07 Boundary Protection
SG.SC-06 Resource Priority
SG.SI-07 Software and Information Integrity
SG.SC-05 Denial-of-Service Protection
SG.SC-08 Communication Integrity

Figure 2-15 Logical Interface Category 12

2.3.9 Logical Interface Category 13: Interface between systems that use the AMI network

Logical interface category 13 covers the interfaces between systems that use the AMI network, for example:

- Between MDMS and meters; and
- Between LMS/DRMS and Customer EMS.

The issues for this interface category include the following:

- Most information from the customer must be treated as confidential.
- Integrity of data is clearly important in general, but alternate means for retrieving and/or validating it can be used.
- Availability is generally low across AMI networks, since they are not designed for real-time interactions or rapid request-response requirements.
- Volume of traffic across AMI networks must be kept low to avoid DoS situations.

- Meters are constrained in their computational capabilities, primarily to keep costs down, which may limit the types and layers of security that could be applied.
- Revenue-grade meters must be certified, so patches and upgrades require extensive testing and validation.
- Meshed wireless communication networks are often used, which can present challenges related to wireless availability as well as throughput and configurations.
- Key management of millions of meters and other equipment will pose significant challenges that have not yet been addressed as standards.
- Remote disconnect could cause unauthorized outages.
- Due to the relatively new technologies used in AMI networks, communication protocols have not yet stabilized as accepted standards, nor have their capabilities been proven through rigorous testing.
- AMI networks span across organizations between utilities with corporate security requirements and customers with no or limited security capabilities or understandings.
- Utility-owned meters are in unsecured locations that are not under utility control, limiting physical security.
- Many possible future interactions across the AMI network are still being designed, are just being speculated about, or have not yet been conceived.
- Customer reactions to AMI systems and capabilities are as yet unknown.

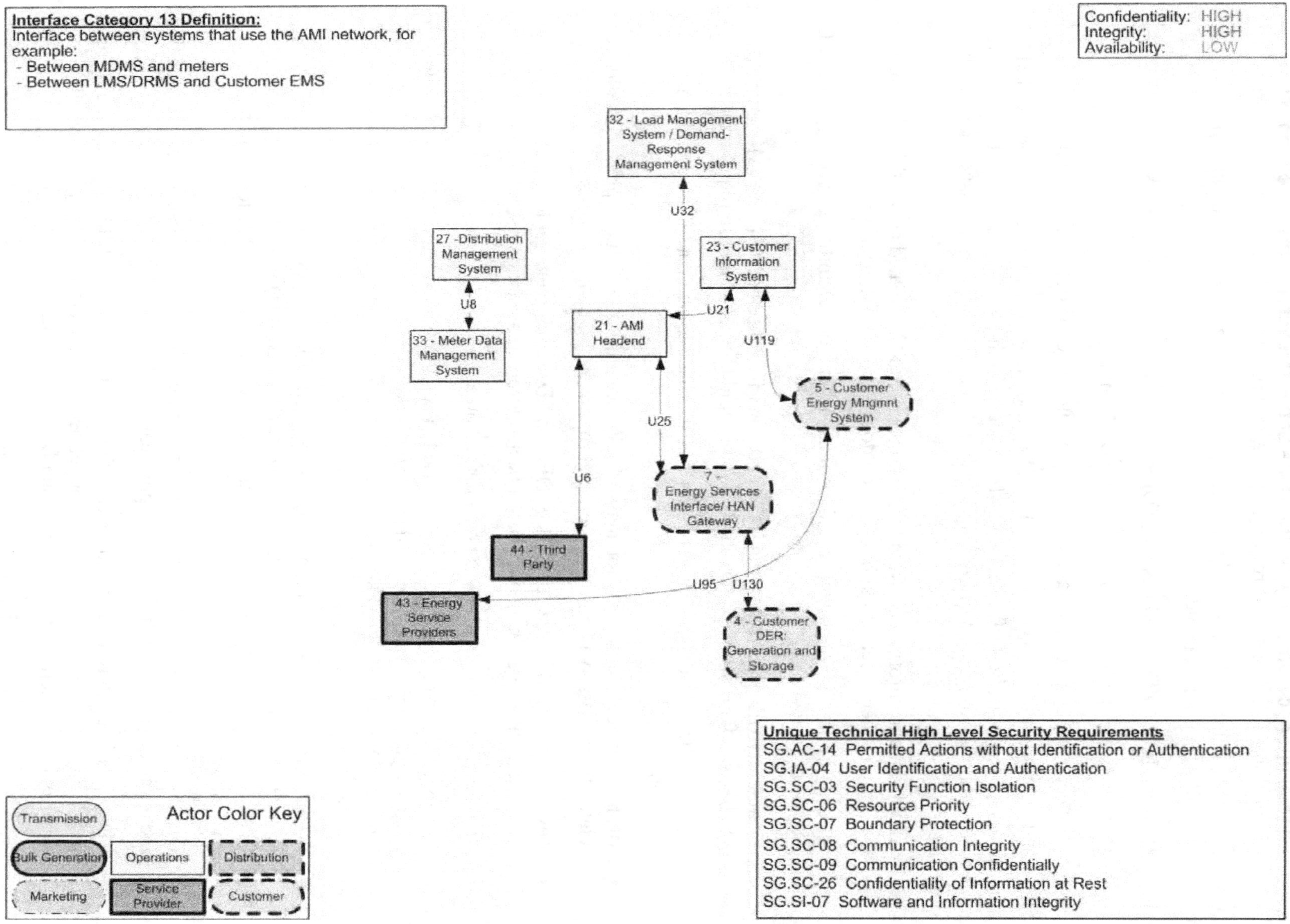

NISTIR 7628 Guidelines for Smart Grid Cyber Security v1.0 – Aug 2010

Interface Category 13 Definition:
Interface between systems that use the AMI network, for example:
- Between MDMS and meters
- Between LMS/DRMS and Customer EMS

Confidentiality: HIGH
Integrity: HIGH
Availability: LOW

Unique Technical High Level Security Requirements
SG.AC-14 Permitted Actions without Identification or Authentication
SG.IA-04 User Identification and Authentication
SG.SC-03 Security Function Isolation
SG.SC-06 Resource Priority
SG.SC-07 Boundary Protection
SG.SC-08 Communication Integrity
SG.SC-09 Communication Confidentially
SG.SC-26 Confidentiality of Information at Rest
SG.SI-07 Software and Information Integrity

Figure 2-16 Logical Interface Category 13

2.3.10 Logical Interface Category 14: Interface between systems that use the AMI network for functions that require high availability

Logical interface category 14 covers the interfaces between systems that use the AMI network with high availability, for example:

- Between LMS/DRMS and customer DER;
- Between DMS applications and customer DER; and
- Between DMS applications and distribution automation (DA) field equipment.

Although both logical interface categories 13 and 14 use the AMI network to connect to field sites, the issues for logical interface category 14 differ from those of 13, because the interactions are focused on power operations of DER and DA equipment. Therefore the issues include the following:

- Although some information from the customer should be treated as confidential, most of the power system operational information does not need to be confidential.
- Integrity of data is very important, since it can affect the reliability and/or efficiency of the power system.
- Availability will need to be a higher requirement for those parts of the AMI networks that will be used for real-time interactions and/or rapid request-response requirements.
- Volume of traffic across AMI networks will still need to be kept low to avoid DoS situations.
- Meshed wireless communication networks are often used, which can present challenges related to wireless availability as well as throughput and configurations.
- Key management of large numbers of DER and DA equipment deployments will pose significant challenges that have not yet been addressed as standards.
- Remote disconnect could cause unauthorized outages.
- Due to the relatively new technologies used in AMI networks, communication protocols have not yet stabilized as accepted standards, nor have their capabilities been proven through rigorous testing. This is particularly true for protocols used for DER and DA interactions.
- AMI networks span across organizations between utilities with corporate security requirements and customers with no or limited security capabilities or understandings. Therefore, maintaining the level of security needed for DER interactions will be challenging.
- DER equipment, and to some degree DA equipment, is found in unsecured locations that are not under utility control, limiting physical security.
- Many possible future interactions across the AMI network are still being designed, are just being speculated about, or have not yet been conceived. These could impact the security of the interactions with DER and DA equipment.

NISTIR 7628 Guidelines for Smart Grid Cyber Security v1.0 – Aug 2010

Interface Category 14 Definition:
Interface between systems that use the AMI network with high availability, for example:
- Between MDMS and meters
- Between LMS/DRMS and Customer EMS
- Between DMS Applications and Customer DER
- Between DMS Applications and DA Field Equipment

Confidentiality: HIGH
Integrity: HIGH
Availability: HIGH

Unique Technical High Level Security Requirements
SG.AC-14 Permitted Actions without Identification or Authentication
SG.IA-04 User Identification and Authentication
SG.SC-03 Security Function Isolation
SG.SC-05 Denial-of-Service Protection
SG.SC-06 Resource Priority
SG.SC-07 Boundary Protection
SG.SC-08 Communication Integrity
SG.SC-09 Communication Confidentiality
SG.SC-26 Confidentiality of Information at Rest
SG.SI-07 Software and Information Integrity

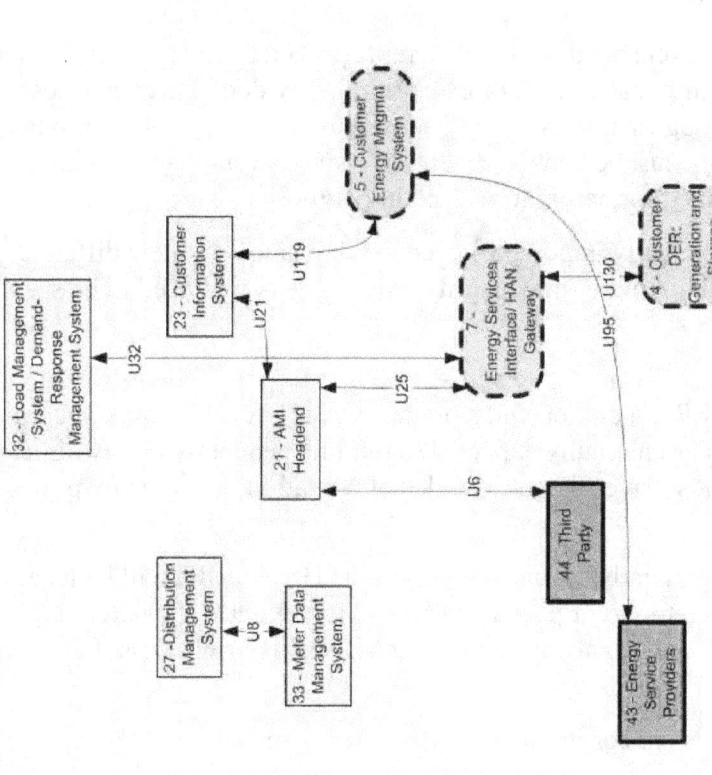

Figure 2-17 Logical Interface Category 14

53

2.3.11 Logical Interface Category 15: Interface between systems that use customer (residential, commercial, and industrial) site networks such as HANs and BANs

Logical interface category 15 covers the interface between systems that use customer (residential, commercial, and industrial) site networks such as home area networks, building/business area networks, and neighborhood area networks (NANs), for example:

- Between customer EMS and customer appliances;
- Between customer EMS and customer DER equipment; and
- Between an energy services interface (ESI) and PEVs.

The security-related issues for this intra-customer site environment HAN/BAN/NAN interface category include the following:

- Some information exchanged among different appliances and systems must be treated as confidential to ensure that an unauthorized third party does not gain access to it. For instance, energy usage statistics from the customer site that are sent through the ESI/HAN gateway must be kept confidential from other appliances whose vendors may want to capture this information for marketing purposes.

- Integrity of data is clearly important in general, but since so many different types of interactions are taking place, the integrity requirements will need to be specific to the particular application.

- Availability is generally moderate across HANs since most interactions are not needed in real time. Even DER generation and storage devices have their own integrated controllers, which are normally expected to run independently of any direct monitoring and control and must have "default" modes of operation to avoid any power system problems.

- Bandwidth is not generally a concern, since most HAN media will be local wireless (e.g., Wi-Fi, ZigBee, Bluetooth) or power line (e.g., HomePlug). The latter may be somewhat bandwidth-limited but can always be replaced by cable or wireless if greater bandwidth is needed.

- Some HAN devices are constrained in their compute capabilities, primarily to keep costs down, which may limit the types and layers of security that could be applied.

- Wireless communication networks are expected to be used within the HAN, which could present some challenges related to wireless configuration and security, because most HANs will not have security experts managing these systems. For instance, if available security measures are not properly set, the HAN security could be compromised by any one of the internal devices, as well as by external entities searching for these insecure HANs.

- Key management of millions of devices within millions of HANs will pose significant challenges that have not yet been addressed as standards.

- Due to the relatively new technologies used in HANs, communication protocols have not yet stabilized as accepted standards, nor have their capabilities been proven through rigorous testing.

- HANs will be accessible by many different vendors and organizations with unknown corporate security requirements and equally variable degrees and types of security solutions. Even if one particular interaction is "secure," in aggregate the multiplicity of interactions may not be secure.

- Some HAN devices may be in unsecured locations, thus limiting physical security. Even those presumably "physically secure" within a home are vulnerable to inadvertent situations such as poor maintenance and misuse, as well as break-ins and theft.

- Many possible future interactions within the HAN environment are still being designed, are just being speculated about, or have not yet been conceived.

NISTIR 7628 Guidelines for Smart Grid Cyber Security v1.0 – Aug 2010

Interface Category 15 Definition:
Interface between systems that use customer (residential, commercial, and industrial) site networks such as HANs and BANs, for example:
- Between Customer EMS and Customer Appliances
- Between Customer EMS and Customer DER
- Between Energy Service Interface and PEV

Confidentiality: LOW
Integrity: MODERATE
Availability: MODERATE

Unique Technical High Level Security Requirements
SG.AC-14 Permitted Actions without Identification or Authentication
SG.IA-04 User Identification and Authentication
SG.SC-03 Security Function Isolation
SG.SC-05 Denial-of-Service Protection
SG.SC-06 Resource Priority
SG.SC-07 Boundary Protection
SG.SC-08 Communication Integrity
SG.SC-09 Communication Confidentially
SG.SC-26 Confidentiality of Information at Rest
SG.SI-07 Software and Information Integrity

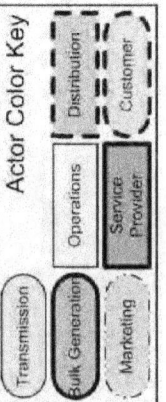

Figure 2-18 Logical Interface Category 15

2.3.12 Logical Interface Category 16: Interface between external systems and the customer site

Logical interface category 16 covers the interface between external systems and the customer site, for example:

- Between a third party and the HAN gateway;
- Between ESP and DER; and
- Between the customer and CIS Web site.

The security-related issues for this external interface to the customer site include the following:

- Some information exchanged among different appliances and systems must be treated as confidential and private to ensure that an unauthorized third party does not gain access to it. For instance, energy usage statistics from the customer site that are sent through the ESI/HAN gateway must be kept confidential from other appliances whose vendors may want to scavenge this information for marketing purposes.

- Integrity of data is clearly important in general, but since so many different types of interactions are taking place, the integrity requirements will need to be specific to the particular application.

- Availability is generally not critical between external parties and the customer site since most interactions are not related to power system operations nor are they needed in real time. Even DER generation and storage devices have their own integrated controllers that are normally expected to run independently of any direct monitoring and control, and should have "default" modes of operation to avoid any power system problems.

- Bandwidth is not generally a concern, since higher-speed media can be used if a function requires a higher volume of data traffic. Many different types of media, particularly public media, are increasingly available, including the public Internet over cable or digital subscriber line (DSL), campus or corporate intranets, cell phone general packet radio service (GPRS), and neighborhood WiMAX and Wi-Fi systems.

- Some customer devices that contain their own "HAN gateway" firewall are constrained in their computational capabilities, primarily to keep costs down, which may limit the types and layers of security which could be applied with those devices.

- Other than those used over the public Internet, communication protocols between third parties and ESI/HAN gateways have not yet stabilized as accepted standards, nor have their capabilities been proven through rigorous testing.

- ESI/HAN gateways will be accessible by many different vendors and organizations with unknown corporate security requirements and equally variable degrees and types of security solutions. Even if one particular interaction is "secure," in aggregate the multiplicity of interactions may not be secure.

- ESI/HAN gateways may be in unsecured locations, thus limiting physical security. Even those presumably "physically secure" within a home are vulnerable to inadvertent situations such as poor maintenance and misuse, as well as break-ins and theft.

- Many possible future interactions within the HAN environment are still being designed, are just being speculated about, or have not yet been conceived, leading to many possible but unknown security issues.

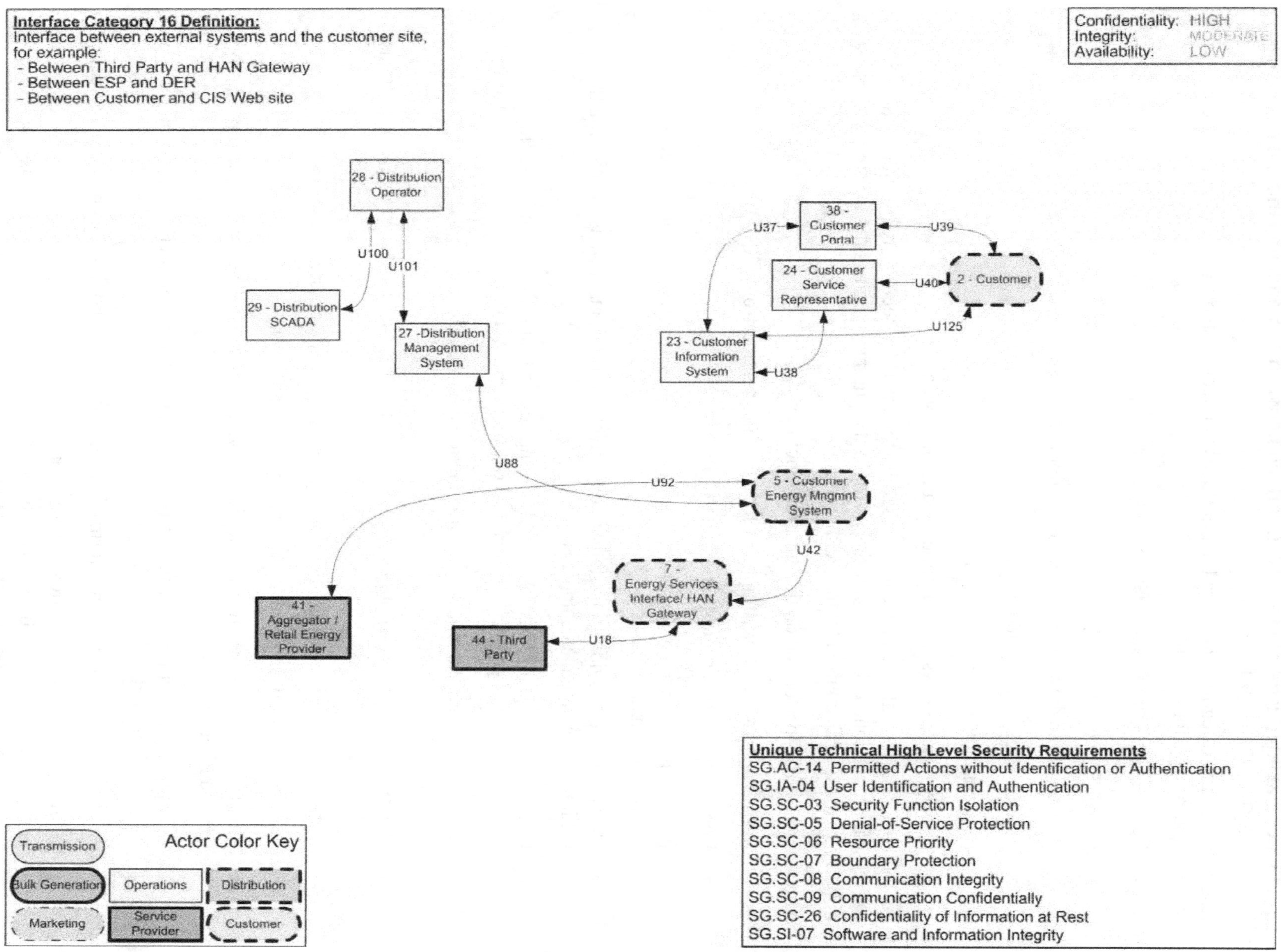

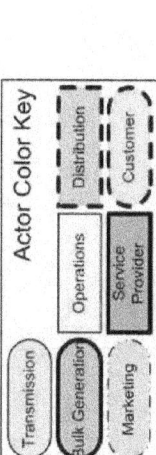

Figure 2-19 Logical Interface Category 16

2.3.13 Logical Interface Category 17: Interface between systems and mobile field crew laptops/equipment

Logical interface category 17 covers the interfaces between systems and mobile field crew laptops/equipment, for example:

- Between field crews and a GIS;
- Between field crews and CIS;
- Between field crews and substation equipment;
- Between field crews and OMS;
- Between field crews and WMS; and
- Between field crews and corporate marketing systems.

As with all other logical interface categories, only the interface security requirements are addressed, not the inherent vulnerabilities of the end equipment such as the laptops or personal digital assistants (PDAs) used by the field crew.

The main activities performed on this interface include:

- Retrieving maps and/or equipment location information from GIS;
- Retrieving customer information from CIS;
- Providing equipment and customer updates, such as meter, payment, and customer information updates to CIS;
- Obtaining and providing substation equipment information, such as location, fault, testing, and maintenance updates;
- Obtaining outage information and providing restoration information, including equipment, materials, and resource information from/to OMS;
- Obtaining project and equipment information and providing project, equipment, materials, resource, and location updates from/to WMS;
- Obtaining metering and outage/restoration verification information from AMI systems; and
- Obtaining customer and product information for upsell opportunities.

The key characteristics of this interface category are as follows:

- This interface is primarily for customer service operations. The most critical needs for this interface are
    - To post restoration information back to the OMS for reprediction of further outage situations; and
    - To receive reconnection information for customers who have been disconnected.
- Information exchanged between these systems is typically corporate-owned, and security is managed within the utility between the interfacing applications. Increased use of wireless technologies and external service providers adds a layer of complexity in

security requirements that is addressed in all areas where multivendor services are interfaced with utility systems.

- Integrity of data is clearly important in general, but since so many different types of interactions are taking place, the integrity requirements will need to be specific to the particular application. However, the integrity of revenue-grade metering data that may be collected in this manner is vital since it has a direct financial impact on all stakeholders of the loads and generation being metered.

- Availability is generally not critical, as interactions are not necessary for real time. Exceptions include payment information for disconnects, restoration operations, and efficiency of resource management.

- Bandwidth is not generally a concern, as most utilities have sized their communications infrastructure to meet the needs of the field applications, and most field applications have been designed for minimal transmission of data in wireless mode. However, more and more applications are being given to field crews to enhance customer service opportunities and for tracking and reporting of construction, maintenance, and outage restoration efforts. This will increase the amount of data and interaction between the corporate systems, third-party providers, and the field crews.

- Data held on laptops and PDAs is vulnerable to physical theft due to the inherent nature of mobile equipment, but those physical security issues will not be addressed in this section. In addition, most mobile field applications are designed to transmit data as it is input, and therefore data is not transmitted when the volume of data is too large to transmit over a wireless connection or when the area does not have wireless coverage. In such cases, data is maintained on the laptop/PDA until it is reconnected to a physical network.

- Note: Data that is captured (e.g., metering data, local device passwords, security parameters) must be protected at the appropriate level.

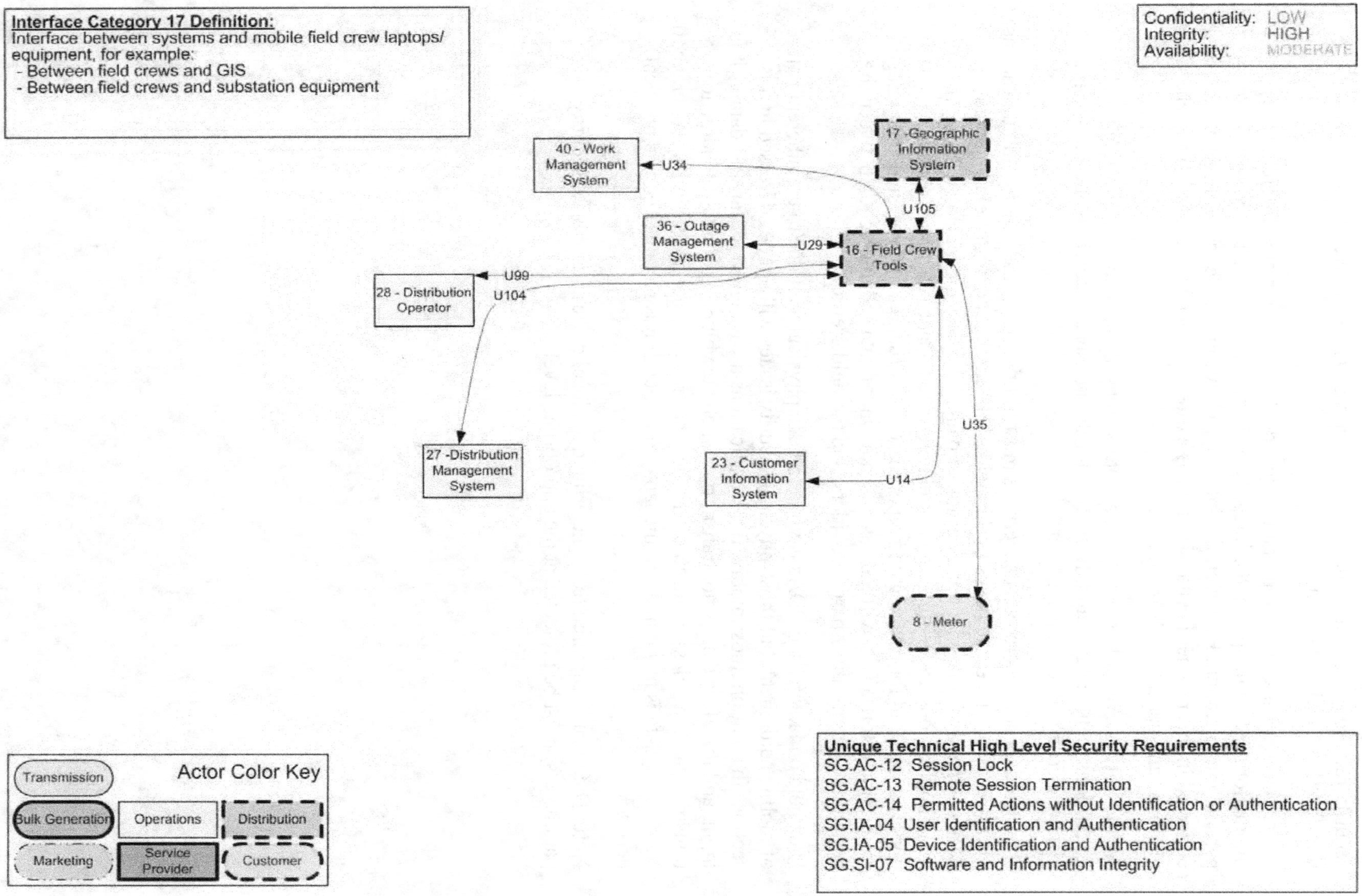
Interface Category 17 Definition:
Interface between systems and mobile field crew laptops/ equipment, for example:
- Between field crews and GIS
- Between field crews and substation equipment
Confidentiality: LOW
Integrity: HIGH
Availability: MODERATE
40 - Work Management System
U34
17 -Geographic Information System
U105
36 - Outage Management System
U29
16 - Field Crew Tools
U99
28 - Distribution Operator
U104
27 -Distribution Management System
U35
23 - Customer Information System
U14
8 - Meter
Actor Color Key
Transmission
Bulk Generation
Operations
Distribution
Marketing
Service Provider
Customer
Unique Technical High Level Security Requirements
SG.AC-12 Session Lock
SG.AC-13 Remote Session Termination
SG.AC-14 Permitted Actions without Identification or Authentication
SG.IA-04 User Identification and Authentication
SG.IA-05 Device Identification and Authentication
SG.SI-07 Software and Information Integrity

Figure 2-20 Logical Interface Category 17

2.3.14 Logical Interface Category 18: Interface between metering equipment

Logical interface category 18 covers the interface between metering equipment, for example:

- Between submeter to meter;
- Between PEV meter and ESP;
- Between MDMS and meters (via the AMI headend);
- Between customer EMS and meters;
- Between field crew tools and meters;
- Between customer DER and submeters; and
- Between electric vehicles and submeters.

The issues for this metering interface category include the following:

- Integrity of revenue grade metering data is vital, since it has a direct financial impact on all stakeholders of the loads and generation being metered.
- Availability of metering data is important but not critical, since alternate means for retrieving metering data can still be used.
- Meters are constrained in their computational capabilities, primarily to keep costs down, which may limit the types and layers of security that could be applied.
- Revenue-grade meters must be certified, so patches and upgrades require extensive testing and validation.
- Key management of millions of meters will pose significant challenges that have not yet been addressed as standards.
- Due to the relatively new technologies used with smart meters, some standards have not been fully developed, nor have their capabilities been proven through rigorous testing.
- Multiple (authorized) stakeholders, including customers, utilities, and third parties, may need access to energy usage either directly from the meter or after it has been processed and validated for settlements and billing, thus adding cross-organizational security concerns.
- Utility-owned meters are in unsecured locations that are not under utility control, limiting physical security.
- Customer reactions to AMI systems and smart meters are as yet unknown.

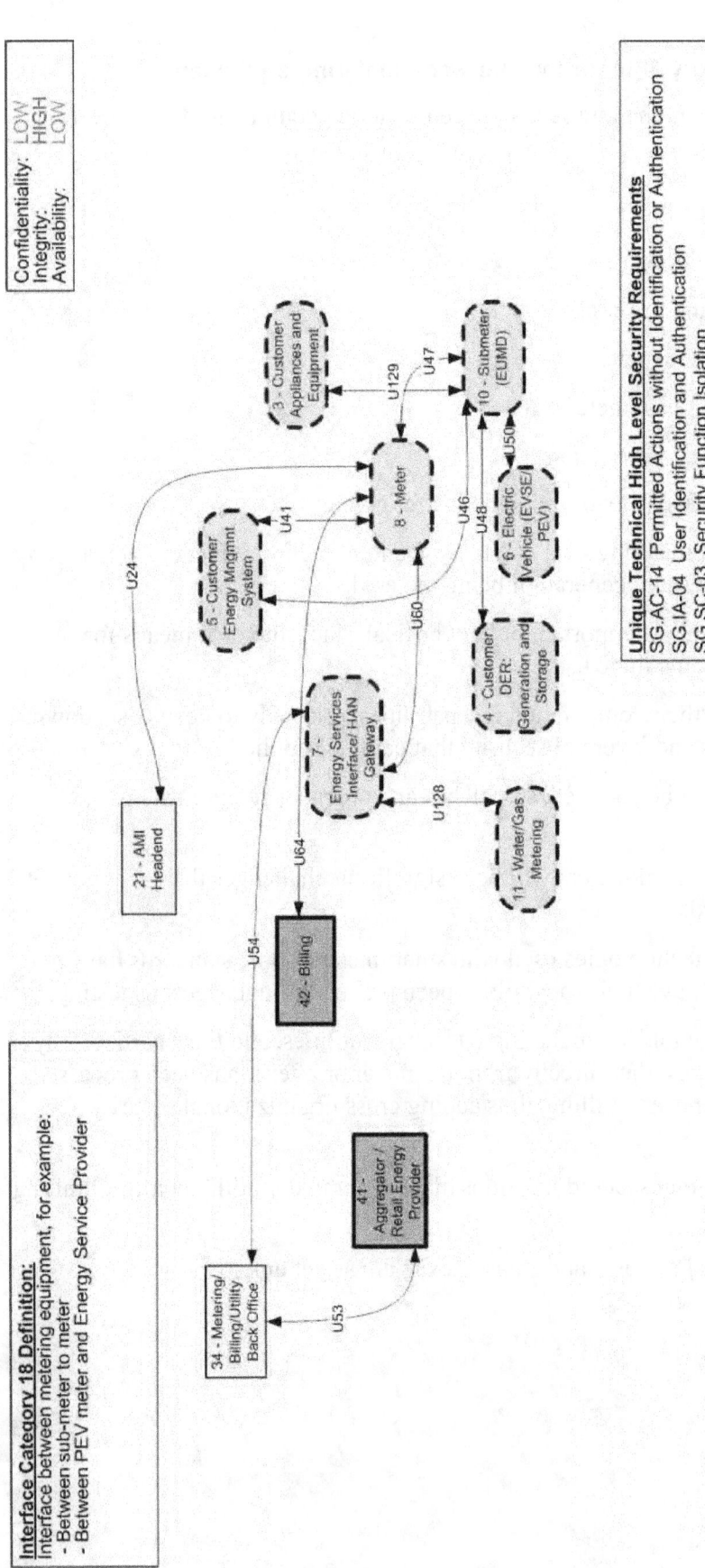

Figure 2-21 Logical Interface Category 18

2.3.15 Logical Interface Category 19: Interface between operations decision support systems

Logical interface category 19 covers the interfaces between operations decision support systems, e.g., between WAMS and ISO/RTOs. Due to the very large coverage of these interfaces, the interfaces are more sensitive to confidentiality requirements than other operational interfaces (see logical interface categories 1-4).

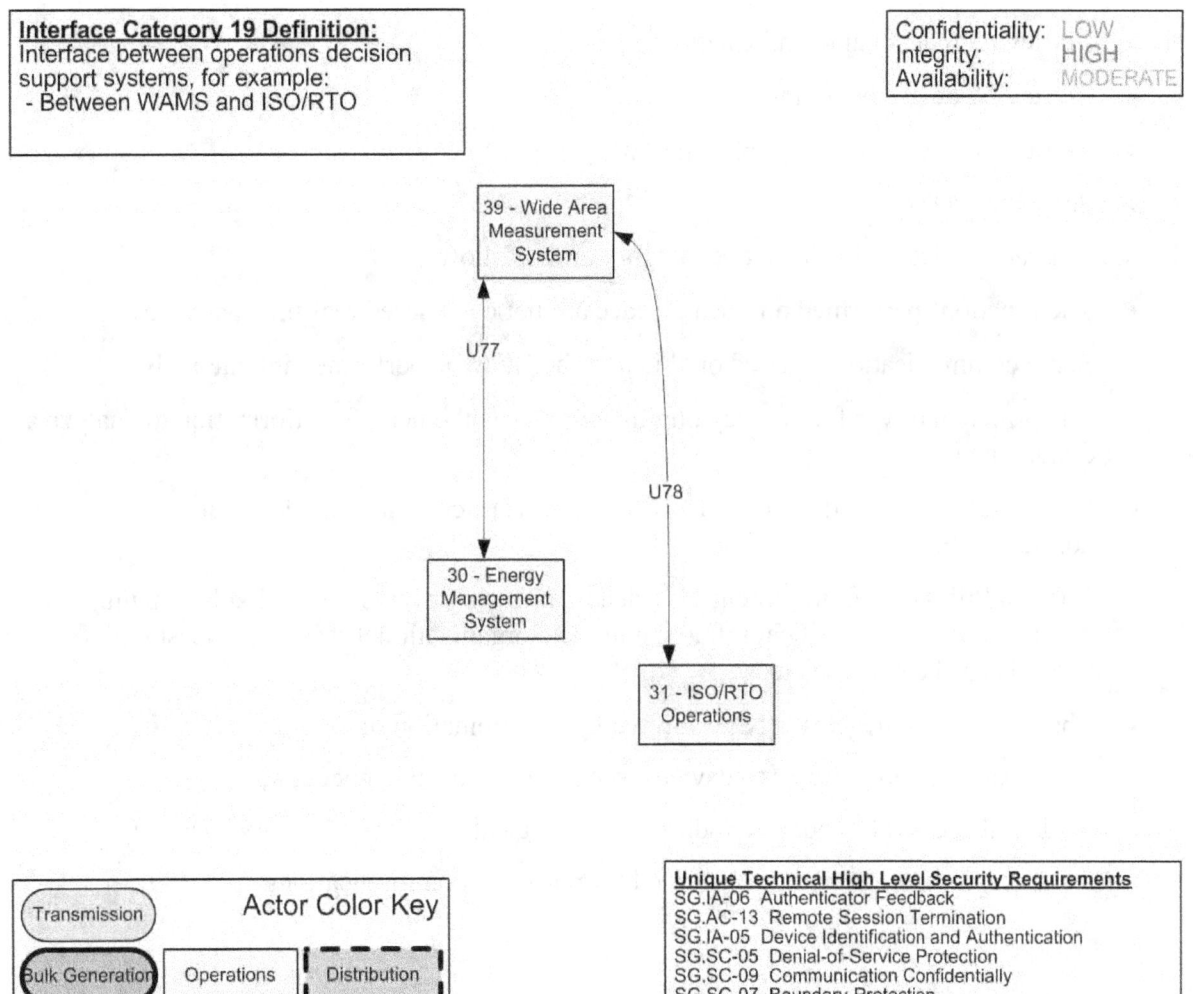

Figure 2-22 Logical Interface Category 19

2.3.16 Logical Interface Category 20: Interface between engineering/ maintenance systems and control equipment

Logical interface category 20 covers the interfaces between engineering/maintenance systems and control equipment, for example:

- Between engineering and substation relaying equipment for relay settings;

- Between engineering and pole-top equipment for maintenance; and
- Within power plants.

The main activities performed on this interface include:

- Installing and changing device settings, which may include operational settings (such as relay settings, thresholds for unsolicited reporting, thresholds for device mode change, and editing of setting groups), event criteria for log record generation, and criteria for oscillography recording;
- Retrieving maintenance information;
- Retrieving device event logs;
- Retrieving device oscillography files; and
- Software updates.

The key characteristics of this interface category are as follows:

- The functions performed on this interface are not considered real-time activities.
- Some communications carried on this interface may be performed interactively.
- The principal driver for urgency on this interface is the need for information to analyze a disturbance.
- Device settings should be treated as critical infrastructure information requiring confidentiality.
- Logs and files containing forensic evidence following events should likely remain confidential for both critical infrastructure and organizational reasons, at least until analysis has been completed.
- These functions are presently performed by a combination of
  - Separate remote access to devices, such as by dial-up connection;
  - Local access at the device (addressed in Logical Interface Category 17); and
  - Access via the same interface used for real-time communications.

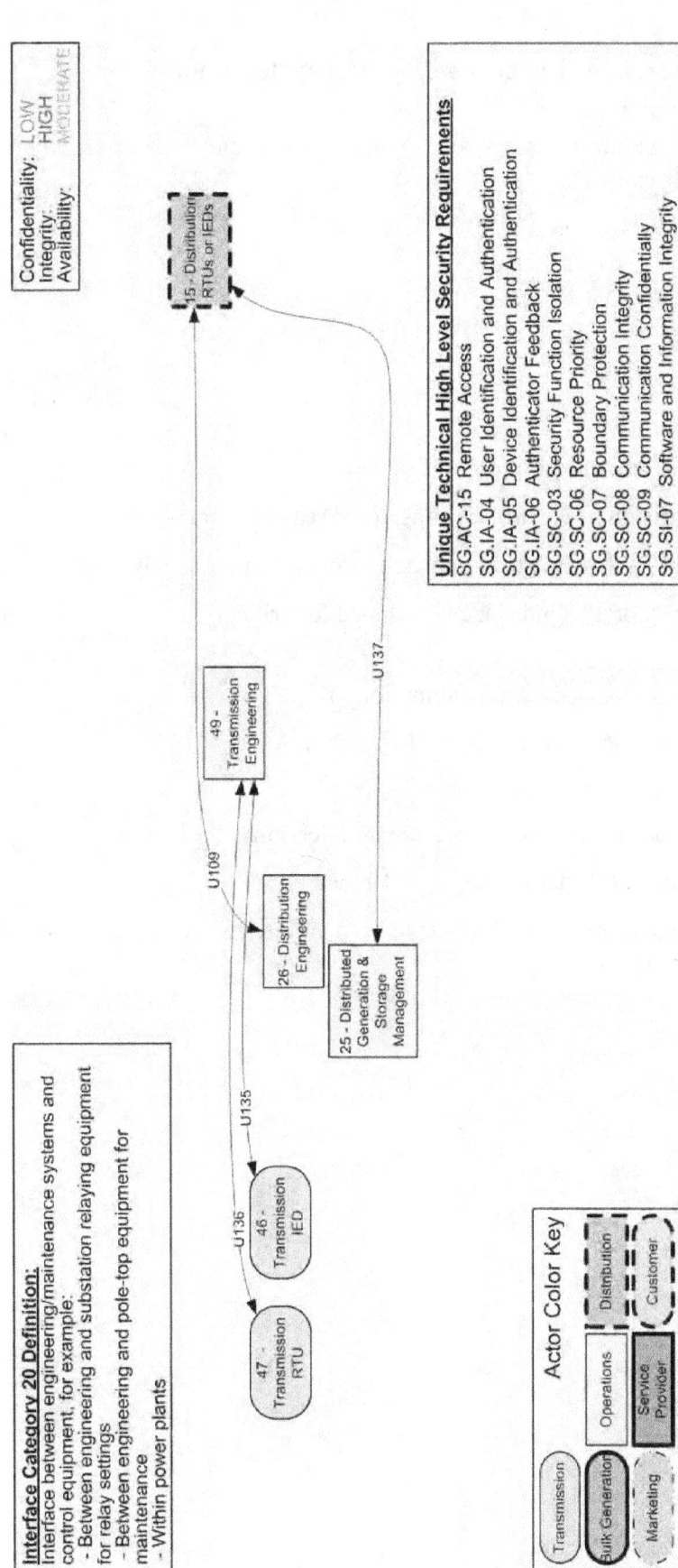

Figure 2-23 Logical Interface Category 20

2.3.17 Logical Interface Category 21: Interface between control systems and their vendors for standard maintenance and service

Logical interface category 21 covers the interfaces between control systems and their vendors for standard maintenance and service, for example:

- Between SCADA system and its vendor.

The main activities performed on this interface include:

- Updating firmware and/or software;
- Retrieving maintenance information; and
- Retrieving event logs.

Key characteristics of this logical interface category are as follows:

- The functions performed on this interface are not considered real-time activities.
- Some communications carried on this interface may be performed interactively.
- The principal driver for urgency on this interface is the need for critical operational/security updates.
- These functions are presently performed by a combination of
    - Separate remote access to devices, such as by dial-up connection;
    - Local access at the device/control system console; and
    - Access via the same interface used for real-time communications.

Activities outside of the scope of Logical Interface Category 21 include:

- Vendors acting in an (outsourced) operational role (see Logical Interface Categories 1-4, 5-6, or 20, depending upon the role).

Interface Category 21 Definition:
Interface between control systems and their vendors for standard maintenance and service, for example:
- Between SCADA system and its vendor

Confidentiality: LOW
Integrity: HIGH
Availability: LOW

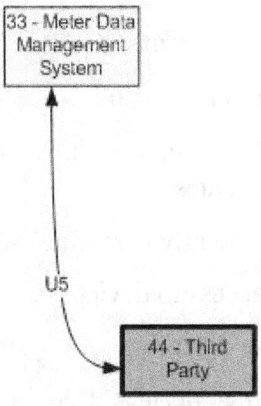

Unique Technical High Level Security Requirements
SG.AC-14 Permitted Actions without Identification or Authentication
SG.AC-15 Remote Access
SG.IA-04 User Identification and Authentication
SG.IA-05 Device Identification and Authentication
SG.IA-06 Authenticator Feedback
SG.SC-03 Security Function Isolation
SG.SC-06 Resource Priority
SG.SC-07 Boundary Protection
SG.SC-08 Communication Integrity
SG.SC-09 Communication Confidentially
SG.SI-07 Software and Information Integrity

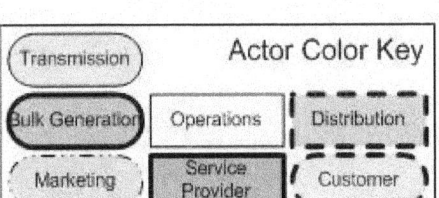

Figure 2-24 Logical Interface Category 21

2.3.18 Logical Interface Category 22: Interface between security/network/system management consoles and all networks and systems

Logical interface category 22 covers the interfaces between security/network/system management consoles and all networks and systems:

- Between a security console and network routers, firewalls, computer systems, and network nodes.

The main activities performed on this interface include:

- Communication infrastructure operations and maintenance;

- Security settings and audit log retrieval (if the security audit log is separate from the event logs);
- Future real-time monitoring of the security infrastructure; and
- Security infrastructure operations and maintenance.

Key characteristics of this logical interface category as follows:

- The functions performed on this interface are not considered real-time activities.
- Some communications carried on this interface may be performed interactively.
- The principal driver for urgency on this interface is the need for critical operational/security updates.
- These functions are presently performed by a combination of
    - Separate remote access to devices, such as by dial-up connection;
    - Local access at the device/control system console; and
    - Access via the same interface used for real-time communications.

Activities outside of the scope of Logical interface category 22 include:

- Smart Grid transmission and distribution (see Logical Interface Categories 1-4 and 5-6);
- Advanced metering (see Logical Interface Category 13); and
- Control systems engineering and systems maintenance (see Logical Interface Category 20).

(Note: This diagram is not included in the logical reference model, Figure 2-3.)

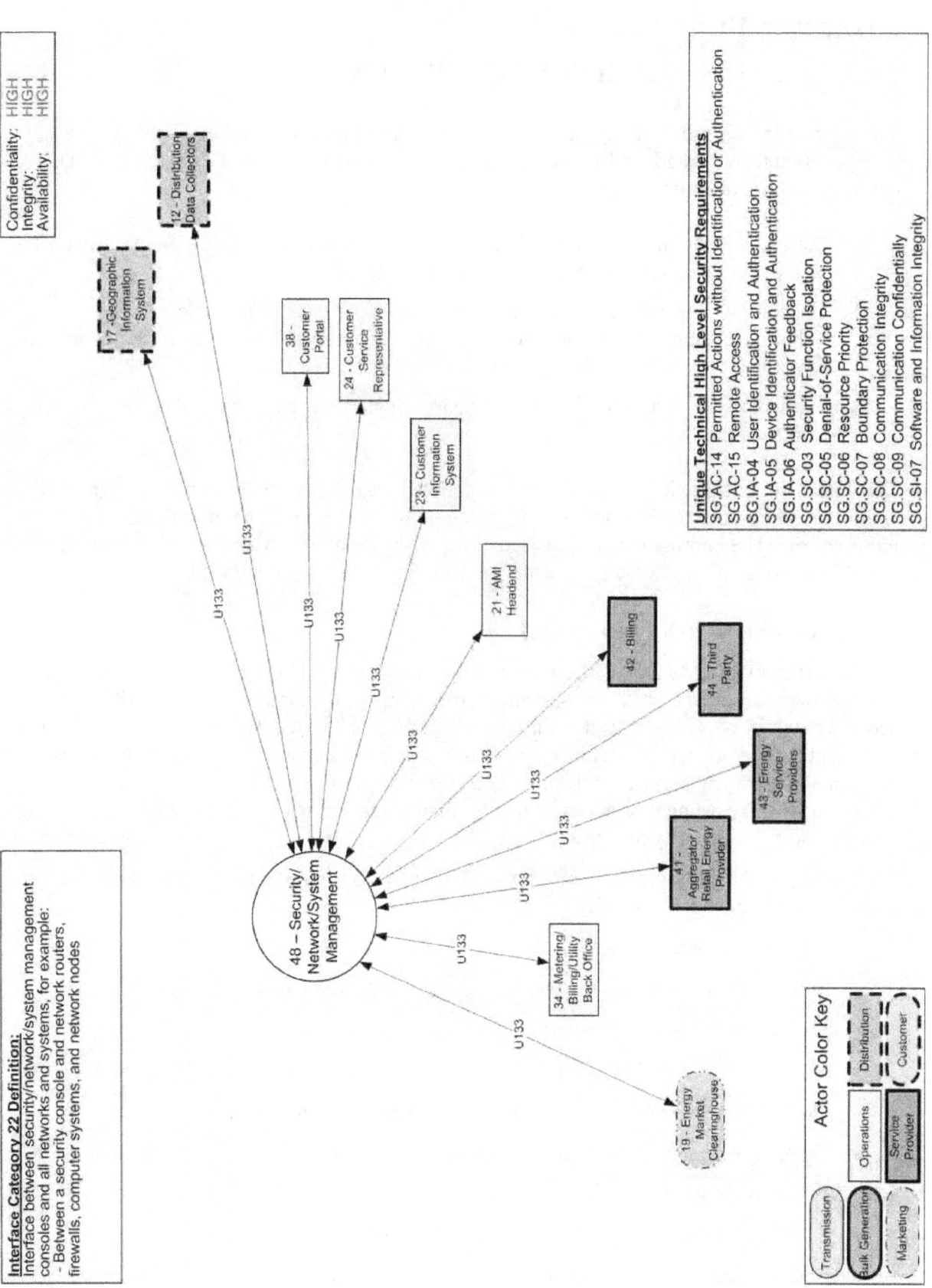

Figure 2-25 Logical Interface Category 22

CHAPTER THREE
HIGH-LEVEL SECURITY REQUIREMENTS

This chapter includes the detailed descriptions for each of the security requirements. The analyses used to select and modify these security requirements are included in Appendix G. This chapter includes the following:

1. Determination of the confidentiality, integrity, and availability (CI&A) impact levels for each of the logical interface categories. (*See* Table 3-2.)
2. The common governance, risk, and compliance (GRC), common technical, and unique technical requirements are allocated to the logical interface categories. Also, the impact levels are included for each requirement. (*See* Table 3-3.)
3. The security requirements for the Smart Grid. Included are the detailed descriptions for each requirement.

This information is provided as guidance to organizations that are implementing, designing, and/or operating Smart Grid systems as a starting point for selecting and modifying security requirements. The information is to be used as a starting point only. Each organization will need to perform a risk analysis to determine the applicability of the following material.

3.1 CYBER SECURITY OBJECTIVES

For decades, power system operations have been managing the reliability of the power grid in which power *availability* has been the primary requirement, with information integrity as a secondary but increasingly critical requirement. Confidentiality of customer information is also important in the normal revenue billing processes and for privacy concerns. Although focused on accidental/inadvertent security problems, such as equipment failures, employee errors, and natural disasters, existing power system management technologies can be used and expanded to provide additional security measures.

Availability is the most important security objective for power system reliability. The time latency associated with availability can vary—

- ≤ 4 ms for protective relaying;
- Subseconds for transmission wide-area situational awareness monitoring;
- Seconds for substation and feeder SCADA data;
- Minutes for monitoring noncritical equipment and some market pricing information;
- Hours for meter reading and longer-term market pricing information; and
- Days/weeks/months for collecting long-term data such as power quality information.

Integrity for power system operations includes assurance that—

- Data has not been modified without authorization;
- Source of data is authenticated;

- Time stamp associated with the data is known and authenticated; and
- Quality of data is known and authenticated.

Confidentiality is the least critical for power system reliability. However, confidentiality is becoming more important, particularly with the increasing availability of customer information online—

- Privacy of customer information;
- Electric market information; and
- General corporate information, such as payroll, internal strategic planning, etc.

3.2 CONFIDENTIALITY, INTEGRITY, AND AVAILABILITY IMPACT LEVELS

Following are the definitions for the security objectives of CI&A, as defined in statute.

Confidentiality

"Preserving authorized restrictions on information access and disclosure, including means for protecting personal privacy and proprietary information…." [44 U.S.C., Sec. 3542]

A loss of *confidentiality* is the unauthorized disclosure of information.

Integrity

"Guarding against improper information modification or destruction, and includes ensuring information non-repudiation and authenticity…." [44 U.S.C., Sec. 3542]

A loss of *integrity* is the unauthorized modification or destruction of information.

Availability

"Ensuring timely and reliable access to and use of information…." [44 U.S.C., SEC. 3542]

A loss of *availability* is the disruption of access to or use of information or an information system.

Based on these definitions, impact levels for each security objective (confidentiality, integrity, and availability) are specified in Table 3-1 as low, moderate, and high as defined in FIPS 199, *Standards for Security Categorization of Federal Information and Information Systems*, February 2004. The impact levels are used in the selection of security requirements for each logical interface category.

Table 3-1 Impact Levels Definitions

	Potential Impact Levels		
	Low	Moderate	High
Confidentiality Preserving authorized restrictions on information access and disclosure, including means for protecting personal privacy and proprietary information. [44 U.S.C., SEC. 3542]	The unauthorized disclosure of information could be expected to have a **limited** adverse effect on organizational operations, organizational assets, or individuals.	The unauthorized disclosure of information could be expected to have a **serious** adverse effect on organizational operations, organizational assets, or individuals.	The unauthorized disclosure of information could be expected to have a **severe or catastrophic** adverse effect on organizational operations, organizational assets, or individuals.
Integrity Guarding against improper information modification or destruction, and includes ensuring information non-repudiation and authenticity. [44 U.S.C., SEC. 3542]	The unauthorized modification or destruction of information could be expected to have a **limited** adverse effect on organizational operations, organizational assets, or individuals.	The unauthorized modification or destruction of information could be expected to have a **serious** adverse effect on organizational operations, organizational assets, or individuals.	The unauthorized modification or destruction of information could be expected to have a **severe or catastrophic** adverse effect on organizational operations, organizational assets, or individuals.
Availability Ensuring timely and reliable access to and use of information. [44 U.S.C., SEC. 3542]	The disruption of access to or use of information or an information system could be expected to have a **limited** adverse effect on organizational operations, organizational assets, or individuals.	The disruption of access to or use of information or an information system could be expected to have a **serious** adverse effect on organizational operations, organizational assets, or individuals.	The disruption of access to or use of information or an information system could be expected to have a **severe or catastrophic** adverse effect on organizational operations, organizational assets, or individuals.

3.3 IMPACT LEVELS FOR THE CI&A CATEGORIES

Each of the three impact levels (i.e., low, moderate, high) is based upon the expected adverse effect of a security breach upon organizational operations, organizational assets, or individuals. The initial designation of impact levels focused on power grid reliability. The expected adverse effect on individuals when privacy breaches occur and adverse effects on financial markets when confidentiality is lost are included here for specific logical interface categories.

Power system reliability: Keep electricity flowing to customers, businesses, and industry. For decades, the power system industry has been developing extensive and sophisticated systems and equipment to avoid or shorten power system outages. In fact, power system operations have been termed the largest and most complex machine in the world. Although there are definitely new areas of cyber security concerns for power system reliability as technology opens new opportunities and challenges, nonetheless, the existing energy management systems and equipment, possibly enhanced and expanded, should remain as key cyber security solutions.

Confidentiality and privacy of customers: As the Smart Grid reaches into homes and businesses, and as customers increasingly participate in managing their energy, confidentiality and privacy of their information has increasingly become a concern. Unlike power system reliability, customer privacy is a new issue.

The impact levels (low [L], moderate [M], and high [H]) presented in Table 3-2 address the impacts to the nationwide power grid, particularly with regard to grid stability and reliability. Consequentially, the confidentiality impact is low for these logical interface categories. Logical interface categories 7, 8, 13, 14, 16, and 22 have a high impact level for confidentiality because of the type of data that needs to be protected (e.g., sensitive customer energy usage data, critical security parameters, and information from a HAN to a third party.)

Table 3-2 Smart Grid Impact Levels

Logical Interface Category	Confidentiality	Integrity	Availability
1	L	H	H
2	L	H	M
3	L	H	H
4	L	H	M
5	L	H	H
6	L	H	M
7	H	M	L
8	H	M	L
9	L	M	M
10	L	H	M
11	L	M	M
12	L	M	M
13	H	H	L
14	H	H	H
15	L	M	M
16	H	M	L
17	L	H	M
18	L	H	L
19	L	H	M
20	L	H	M
21	L	H	L
22	H	H	H

3.4 SELECTION OF SECURITY REQUIREMENTS

Power system operations pose many security challenges that are different from most other industries. For example, the Internet is different from the power system operations environment. In particular, there are strict performance and reliability requirements that are needed by power system operations. For instance—

- Operation of the power system must continue 24×7 with high availability (e.g., 99.99% for SCADA and higher for protective relaying) regardless of any compromise in security or the implementation of security measures that hinder normal or emergency power system operations.

- Power system operations must be able to continue during any security attack or compromise (as much as possible).

- Power system operations must recover quickly after a security attack or the compromise of an information system.

- Testing of security measures cannot be allowed to impact power system operations.

There is no single set of cyber security requirements that addresses each of the Smart Grid logical interface categories. This information can be used as guidelines for organizations as they develop their cyber security strategy, perform risk assessments, and select and modify security requirements for Smart Grid information system implementations.

Additional criteria must be used in determining the cyber security requirements before selecting and implementing the cyber security measures/solutions. These additional criteria must take into account the characteristics of the interface, including the constraints and issues posed by device and network technologies, the existence of legacy components/devices, varying organizational structures, regulatory and legal policies, and cost criteria.

Once these interface characteristics are applied, then cyber security requirements can be applied that are both specific enough to be applicable to the interfaces and general enough to permit the implementation of different cyber security solutions that meet the security requirements or embrace new security technologies as they are developed. This cyber security information can then be used in subsequent steps to select security requirements for the Smart Grid.

The security requirements listed below are an amalgam from several sources: NIST SP 800-53, the DHS Catalog, NERC CIPs, and the NRC Regulatory Guidance. After the security requirements were selected, they were modified as required. The goal was to develop a set of security requirements that address the needs of the electric sector and the Smart Grid. Each security requirement is allocated to one of three categories: governance, risk, and compliance (GRC), common technical, or unique technical. The intent of the GRC requirements is to have them addressed at the organization level. It may be necessary to augment these organization-level requirements for specific logical interface categories and/or Smart Grid information systems. The common technical requirements are applicable to all of the logical interface categories. The unique technical requirements are allocated to one or more of the logical interface categories. The common and unique technical requirements should be allocated to each Smart Grid system and not necessarily to every component within a system, as the focus is on security at the system level. Each organization must develop a security architecture for each Smart Grid information system and allocate security requirements to components/devices. Some security requirements may be allocated to one or more components/devices. However, not every security requirement must be allocated to every component/device. Table 3-3 includes only the security requirements that were selected. There are additional security requirements included in the next section that were not selected. These may be included by an organization if it determines that the security requirements are necessary to address specific risks and needs.

For each unique technical requirement, the recommended security impact level is specified (e.g., low [L], moderate [M], or high [H]). The common technical requirements and GRC requirements apply to all logical interface categories. A recommended impact level is included with each of the common technical and GRC requirements. The requirement may be the same at all impact levels. If there are additional requirements at the moderate and high impact levels, these are listed in the table. The information included in the table is a guideline and presented as a starting point for organizations as they implement Smart Grid information systems. Each organization should use this guidance information as it implements the security strategy and performs the security risk assessment.

In addition, organizations may find it necessary to identify compensating security requirements. A compensating security requirement is implemented by an organization in lieu of a recommended security requirement to provide equivalent or comparable level of protection for the information/control system and the information processed, stored, or transmitted by that system. More than one compensating requirement may be required to provide the equivalent or comparable protection for a particular security requirement. For example, an organization with significant staff limitations may compensate for the recommended separation of duty security requirement by strengthening the audit, accountability, and personnel security requirements within the information/control system.

3.5 SECURITY REQUIREMENTS EXAMPLE

This example illustrates how to select security requirements using the material in this report. Included in this example are some GRC, common technical and unique technical requirements that may apply to a Smart Grid information system.

Example: Smart Grid control system "ABC" includes logical interface category 6: interface between control systems in different organizations. As specified in the previous chapter, this requires high data accuracy, high availability, and establishment of a chain of trust.

The organization will need to review all the GRC requirements to determine if any of these requirements need to be modified or augmented for the ABC control system. For example, SG.AC-1, Access Control Policy and Procedures, is applicable to all systems, including the ABC control system. This security requirement does not need to be revised for the ABC control system because it is applicable at the organization level. In contrast, for GRC requirement SG.CM-6, Configuration Settings, the organization determines that there are unique settings for the ABC control system.

For common technical requirement SG.SI-2, Flaw Remediation, the organization determines that the procedures already specified are applicable to the ABC control system, without modification. In contrast, for common technical requirement SG.AC-7, Least Privilege, the organization determines that a unique set of access rights and privileges are necessary for the ABC control system because the system interconnects with a system in a different organization.

Unique technical requirement SG.SI-7, Software and Information Integrity, was allocated to logical interface category 6. The organization has determined that this security requirement is important for the ABC control system, and includes it in the suite of security requirements.

3.6 RECOMMENDED SECURITY REQUIREMENTS

Table 3-3 lists the selected security requirements for the Smart Grid.

Table 3-3 Allocation of Security Requirements to Logical Interface Categories

Dark Gray = Unique Technical Requirement Light Gray = Common Technical Requirement
White = Common Governance, Risk and Compliance (GRC)

Smart Grid Requirement Number	Logical Interface Categories																					
	1	2	3	4	5	6	7	8	9	10	11	12	13	14	15	16	17	18	19	20	21	22
SG.AC-1	Applies at all impact levels																					
SG.AC-2	Applies at all impact levels																					
SG.AC-3	Applies at all impact levels																					
SG.AC-4	Applies at all impact levels																					
SG.AC-6	Applies at moderate and high impact levels																					
SG.AC-7	Applies at moderate and high impact levels																					
SG.AC-8	Applies at all impact levels																					
SG.AC-9	Applies at all impact levels																					
SG.AC-12							H	H									L				L	H
SG.AC-13																	M		M			
SG.AC-14	H	H	H	H	H	H	M	M	M	H			H	H	M	M	H	H	H	H	H	H
SG.AC-15																				H	H	H
SC.AC-16	Applies at all impact levels																					
SG.AC-17	Applies at all impact levels with additional requirement enhancements at moderate and high impact levels																					
SG.AC-18	Applies at all impact levels with additional requirement enhancements at moderate and high impact levels																					
SG.AC-19	Applies at all impact levels																					
SG.AC-20	Applies at all impact levels																					
SG.AC-21	Applies at all impact levels																					
SG.AT-1	Applies at all impact levels																					

NISTIR 7628 Guidelines for Smart Grid Cyber Security v1.0 – Aug 2010

Dark Gray = Unique Technical Requirement   Light Gray = Common Technical Requirement
White = Common Governance, Risk and Compliance (GRC)

Smart Grid Requirement Number	Logical Interface Categories																					
	1	2	3	4	5	6	7	8	9	10	11	12	13	14	15	16	17	18	19	20	21	22
SG.AT-2	Applies at all impact levels
SG.AT-3	Applies at all impact levels
SG.AT-4	Applies at all impact levels
SG.AT-6	Applies at all impact levels
SG.AT-7	Applies at all impact levels
SG.AU-1	Applies at all impact levels
SG.AU-2	Applies at all impact levels with additional requirement enhancements at high impact level
SG.AU-3	Applies at all impact levels
SG.AU-4	Applies at all impact levels
SG.AU-5	Applies at all impact levels with additional requirement enhancements at high impact level
SG.AU-6	Applies at all impact levels
SG.AU-7	Applies at moderate and high impact levels
SG.AU-8	Applies at all impact levels with additional requirement enhancements at moderate and high impact levels
SG.AU-9	Applies at all impact levels
SG.AU-10	Applies at all impact levels
SG.AU-11	Applies at all impact levels
SG.AU-12	Applies at all impact levels
SG.AU-13	Applies at all impact levels
SG.AU-14	Applies at all impact levels
SG.AU-15	Applies at all impact levels

Dark Gray = Unique Technical Requirement   Light Gray = Common Technical Requirement
White = Common Governance, Risk and Compliance (GRC)

Smart Grid Requirement Number	Logical Interface Categories																					
	1	2	3	4	5	6	7	8	9	10	11	12	13	14	15	16	17	18	19	20	21	22
SG.AU-16							M	M	M				H	H		M				H	H	H
SG.CA-1	Applies at all impact levels
SG.CA-2	Applies at all impact levels
SG.CA-4	Applies at all impact levels
SG.CA-5	Applies at all impact levels
SG.CA-6	Applies at all impact levels
SG.CM-1	Applies at all impact levels
SG.CM-2	Applies at all impact levels
SG.CM-3	Applies at moderate and high impact levels
SG.CM-4	Applies at all impact levels
SG.CM-5	Applies at moderate and high impact levels
SG.CM-6	Applies at all impact levels
SG.CM-7	Applies at all impact levels
SG.CM-8	Applies at all impact levels
SG.CM-9	Applies at all impact levels
SG.CM-10	Applies at all impact levels
SG.CM-11	Applies at all impact levels
SG.CP-1	Applies at all impact levels

Dark Gray = Unique Technical Requirement Light Gray = Common Technical Requirement
White = Common Governance, Risk and Compliance (GRC)

Smart Grid Requirement Number	Logical Interface Categories																						
	1	2	3	4	5	6	7	8	9	10	11	12	13	14	15	16	17	18	19	20	21	22
SG.CP-2	Applies at all impact levels																						
SG.CP-3	Applies at all impact levels																						
SG.CP-4	Applies at all impact levels																						
SG.CP-5	Applies at moderate and high impact levels																						
SG.CP-6	Applies at all impact levels																						
SG.CP-7	Applies at moderate and high impact levels with additional requirement enhancements at moderate and high impact levels																						
SG.CP-8	Applies at moderate and high impact levels with additional requirement enhancements at moderate and high impact levels																						
SG.CP-9	Applies at moderate and high impact levels with additional requirement enhancements at moderate and high impact levels																						
SG.CP-10	Applies at all impact levels with additional requirement enhancements at moderate and high impact levels																						
SG.CP-11	Applies at high impact levels																						
SG.IA-1	Applies at all impact levels																						
SG.IA-2	Applies at all impact levels																						
SG.IA-3	Applies at all impact levels																						
SG.IA-4	H	H	H	H	H	H	M	M	M	H			H	H	M	M	H	H	H	H	H	H
SG.IA-5	H	H	H	H	H	M	M	M	H		M			H	H	H	H	H	H	H
SG.IA-6	L	L	L	L	L	L	H	H	L	L			H	H	H	L	L	L	L	L	L	H
SG.ID-1	Applies at all impact levels																						
SG.ID-2	Applies at all impact levels																						

Dark Gray = Unique Technical Requirement
Light Gray = Common Technical Requirement
White = Common Governance, Risk and Compliance (GRC)

Smart Grid Requirement Number	Logical Interface Categories																					
	1	2	3	4	5	6	7	8	9	10	11	12	13	14	15	16	17	18	19	20	21	22
SG.ID-3	Applies at all impact levels																					
SG.ID-4	Applies at all impact levels																					
SG.IR-1	Applies at all impact levels																					
SG.IR-2	Applies at all impact levels																					
SG.IR-3	Applies at all impact levels																					
SG.IR-4	Applies at all impact levels																					
SG.IR-5	Applies at all impact levels																					
SG.IR-6	Applies at all impact levels																					
SG.IR-7	Applies at all impact levels																					
SG.IR-8	Applies at all impact levels																					
SG.IR-9	Applies at all impact levels																					
SG.IR-10	Applies at all impact levels with additional requirement enhancements at moderate and high impact levels																					
SG.IR-11	Applies at all impact levels																					
SG.MA-1	Applies at all impact levels																					
SG.MA-2	Applies at all impact levels																					
SG.MA-3	Applies at all impact levels with additional requirement enhancements at high impact levels																					
SG.MA-4	Applies at all impact levels																					
SG.MA-5	Applies at all impact levels																					
SG.MA-6	Applies at all impact levels with additional requirement enhancements at high impact levels																					

NISTIR 7628 Guidelines for Smart Grid Cyber Security v1.0 – Aug 2010

Dark Gray = Unique Technical Requirement   Light Gray = Common Technical Requirement
White = Common Governance, Risk and Compliance (GRC)

Smart Grid Requirement Number	Logical Interface Categories																					
	1	2	3	4	5	6	7	8	9	10	11	12	13	14	15	16	17	18	19	20	21	22
SG.MA-7	Applies at all impact levels																					
SG.MP-1	Applies at all impact levels																					
SG.MP-2	Applies at all impact levels																					
SG.MP-3	Applies at moderate and high impact levels																					
SG.MP-4	Applies at all impact levels																					
SG.MP-5	Applies at all impact levels																					
SG.MP-6	Applies at all impact levels with additional requirement enhancements at moderate and high impact levels																					
SG.PE-1	Applies at all impact levels																					
SG.PE-2	Applies at all impact levels																					
SG.PE-3	Applies at all impact levels with additional requirement enhancements at moderate and high impact levels																					
SG.PE-4	Applies at all impact levels																					
SG.PE-5	Applies at all impact levels with additional requirement enhancements at moderate and high impact levels																					
SG.PE-6	Applies at all impact levels																					
SG.PE-7	Applies at all impact levels																					
SG.PE-8	Applies at all impact levels																					
SG.PE-9	Applies at all impact levels with additional requirement enhancements at moderate and high impact levels																					
SG.PE-10	Applies at all impact levels																					
SG.PE-11	Applies at all impact levels																					
SG.PE-12	Applies at all impact levels with additional requirement enhancements at high impact level																					

Dark Gray = Unique Technical Requirement   Light Gray = Common Technical Requirement
White = Common Governance, Risk and Compliance (GRC)

Smart Grid Requirement Number	Logical Interface Categories																					
	1	2	3	4	5	6	7	8	9	10	11	12	13	14	15	16	17	18	19	20	21	22
SG.PL-1	Applies at all impact levels																				
SG.PL-2	Applies at all impact levels																				
SG.PL-3	Applies at all impact levels																				
SG.PL-4	Applies at all impact levels																				
SG.PL-5	Applies at all impact levels																				
SG.PM-1	Applies at all impact levels																				
SG.PM-2	Applies at all impact levels																				
SG.PM-3	Applies at all impact levels																				
SG.PM-4	Applies at all impact levels																				
SG.PM-5	Applies at all impact levels																				
SG.PM-6	Applies at all impact levels																				
SG.PM-7	Applies at all impact levels																				
SG.PM-8	Applies at all impact levels																				
SG.PS-1	Applies at all impact levels																				
SG.PS-2	Applies at all impact levels																				
SG.PS-3	Applies at all impact levels																				
SG.PS-4	Applies at all impact levels																				
SG.PS-5	Applies at all impact levels																				
SG.PS-6	Applies at all impact levels																				

Dark Gray = Unique Technical Requirement   Light Gray = Common Technical Requirement
White = Common Governance, Risk and Compliance (GRC)

Smart Grid Requirement Number	Logical Interface Categories																					
	1	2	3	4	5	6	7	8	9	10	11	12	13	14	15	16	17	18	19	20	21	22
SG.PS-7	Applies at all impact levels																					
SG.PS-8	Applies at all impact levels																					
SG.PS-9	Applies at all impact levels																					
SG.RA-1	Applies at all impact levels																					
SG.RA-2	Applies at all impact levels																					
SG.RA-3	Applies at all impact levels																					
SG.RA-4	Applies at all impact levels																					
SG.RA-5	Applies at all impact levels																					
SG.RA-6	Applies at all impact levels with additional requirement enhancements at moderate and high impact levels																					
SG.SA-1	Applies at all impact levels																					
SG.SA-2	Applies at all impact levels																					
SG.SA-3	Applies at all impact levels																					
SG.SA-4	Applies at all impact levels																					
SG.SA-5	Applies at all impact levels																					
SG.SA-6	Applies at all impact levels																					
SG.SA-7	Applies at all impact levels																					
SG.SA-8	Applies at all impact levels																					
SG.SA-9	Applies at all impact levels																					
SG.SA-10	Applies at all impact levels																					

NISTIR 7628 Guidelines for Smart Grid Cyber Security v1.0 – Aug 2010

Dark Gray = Unique Technical Requirement Light Gray = Common Technical Requirement
White = Common Governance, Risk and Compliance (GRC)

Smart Grid Requirement Number	Logical Interface Categories																					
	1	2	3	4	5	6	7	8	9	10	11	12	13	14	15	16	17	18	19	20	21	22
SG.SA-11	Applies at all impact levels																					
SG.SC-1	Applies at all impact levels																					
SG.SC-3	H	H	H	H			M						H	H	M	M		H		H	H	H
SG.SC-5	H	M	H	M	M	M			M	M		M		H	M	M			M			H
SG.SC-6					H									H				H				H
SG.SC-7	H	H	H	H	H	H		M	M	H		M	H	H	M	M		H	H	H	H	H
SG.SC-8	H	H	H	H	H	H	M	M	M	H	M	M	H	H	M	M		H	H	H	H	H
SG.SC-9	H	H											H	H		H						H
SG.SC-11	Applies at all impact levels with additional requirement enhancements at high impact levels																					
SG.SC-12	Applies at all impact levels																					
SG.SC-13	Applies at all impact levels																					
SG.SC-15	Applies at all impact levels																					
SG.SC-16	Applies at moderate and high impact levels																					
SG.SC-18	Applies at all impact levels																					
SG.SC-19	Applies at all impact levels																					
SG.SC-20	Applies at all impact levels																					
SG.SC-21	Applies at all impact levels																					
SG.SC-22	Applies at moderate and high impact levels																					
SG.SC-26							H	H		H			H	H		H					H	H
SG.SC-29	H	H	H	H	H	H	H	H		H			H	H			H	H	H	H	H	H

Dark Gray = Unique Technical Requirement   Light Gray = Common Technical Requirement
White = Common Governance, Risk and Compliance (GRC)

Smart Grid Requirement Number	Logical Interface Categories																						
	1	2	3	4	5	6	7	8	9	10	11	12	13	14	15	16	17	18	19	20	21	22
SG.SC-30	Applies at moderate and high impact levels																					
SG.SI-1	Applies at all impact levels																					
SG.SI-2	Applies at all impact levels																					
SG.SI-3	Applies at all impact levels																					
SG.SI-4	Applies at all impact levels																					
SG.SI-5	Applies at all impact levels																					
SG.SI-6	Applies at moderate and high impact levels																					
SG.SI-7	H	H	H	H	H	H	H	M	M	H		M	H	H	M	M	H	H	H	H	H	
SG.SI-8	Applies at moderate and high impact levels																					
SG.SI-9	Applies at all impact levels																					

3.6.1 Security Requirements

This section contains the recommended security requirements for the Smart Grid. The recommended security requirements are organized into families primarily based on NIST SP 800-53. A cross-reference of the Smart Grid security requirements to NIST SP 800-53, the DHS Catalog, and the NERC CIPs is included in Appendix A.

The following information is included with each security requirement:

1. Security requirement identifier and name. Each security requirement has a unique identifier that consists of three components. The initial component is SG – for Smart Grid. The second component is the family name, e.g., AC for access control and CP for Continuity of Operations. The third component is a unique numeric identifier, for example, SG.AC-1 and SG.CP-3. Each requirement also has a unique name.

2. Category. Identifies whether the security requirement is a GRC, common technical, or unique technical requirement. For each common technical security requirement, the most applicable objective (confidentiality, integrity, and availability) is listed.

3. The *Requirement* describes specific security-related activities or actions to be carried out by the organization or by the Smart Grid information system.

4. The *Supplemental Guidance* section provides additional information that may be useful in understanding the security requirement. This information is guidance and is not part of the security requirement.

5. The *Requirement Enhancements* provide statements of security capability to (i) build additional functionality in a requirement, and/or (ii) increase the strength of a requirement. In both cases, the requirement enhancements are used in a Smart Grid information system requiring greater protection due to the potential impact of loss based on the results of a risk assessment. Requirement enhancements are numbered sequentially within each requirement.

6. The *Additional Considerations* provide additional statements of security capability that may be used to enhance the associated security requirement. These are provided for organizations to consider as they implement Smart Grid information systems and are not intended as security requirements. Each additional consideration is number A1, A2, etc., to distinguish them from the security requirements and requirement enhancements.

7. The *Impact Level Allocation* identifies the security requirement and requirement enhancements, as applicable, at each impact level: low, moderate, and high. The impact levels for a specific Smart Grid information system will be determined by the organization in the risk assessment process.

The term *information* is used to include data that is received and data that is sent—including, for example, data that is interpreted as a command, a setting, or a request to send data.

The requirements related to emergency lighting, fire protection, temperature and humidity controls, water damage, power equipment and power cabling, and lockout/tagout[21] are important

---

[21] Lockout/tagout is a safety procedure which is used in industry to ensure that dangerous machines are properly shut off and not started up again prior to the completion of maintenance or servicing work.

requirements for safety. These are outside the scope of cyber security and are not included in this report. However, these requirements must be addressed by each organization in accordance with local, state, federal, and organizational regulations, policies, and procedures.

The requirements related to privacy are not included in this chapter. They are included in Chapter 5 of this report. Specifically, privacy principle recommendations based on the PIA are included in §5.4.2, Summary PIA Findings and Recommendations, and in §5.8, Smart Grid Privacy Summary and Recommendations.

3.7 ACCESS CONTROL (SG.AC)

The focus of access control is ensuring that resources are accessed only by the appropriate personnel, and that personnel are correctly identified. Mechanisms need to be in place to monitor access activities for inappropriate activity.

SG.AC-1 Access Control Policy and Procedures

Category: Common Governance, Risk, and Compliance (GRC) Requirements

Requirement

1. The organization develops, implements, reviews, and updates on an organization-defined frequency—
   a. A documented access control security policy that addresses—
      i. The objectives, roles, and responsibilities for the access control security program as it relates to protecting the organization's personnel and assets; and
      ii. The scope of the access control security program as it applies to all of the organizational staff, contractors, and third parties.
   b. Procedures to address the implementation of the access control security policy and associated access control protection requirements.
2. Management commitment ensures compliance with the organization's security policy and other regulatory requirements; and
3. The organization ensures that the access control security policy and procedures comply with applicable federal, state, local, tribal, and territorial laws and regulations.

Supplemental Guidance

The access control policy can be included as part of the general information security policy for the organization. Access control procedures can be developed for the security program in general and for a particular Smart Grid information system when required.

Requirement Enhancements

None.

Additional Considerations

None.

Impact Level Allocation

| Low: SG.AC-1 | Moderate: SG.AC-1 | High: SG.AC-1 |

SG.AC-2 Remote Access Policy and Procedures

Category: Common Governance, Risk, and Compliance (GRC) Requirements

Requirement

The organization—

1. Documents allowed methods of remote access to the Smart Grid information system;
2. Establishes usage restrictions and implementation guidance for each allowed remote access method;
3. Authorizes remote access to the Smart Grid information system prior to connection; and
4. Enforces requirements for remote connections to the Smart Grid information system.

Supplemental Guidance

Remote access is any access to an organizational Smart Grid information system by a user (or process acting on behalf of a user) communicating through an external, non-organization-controlled network (e.g., the Internet).

Requirement Enhancements

None.

Additional Considerations

None.

Impact Level Allocation

| Low: SG.AC-2 | Moderate: SG.AC-2 | High: SG.AC-2 |

SG.AC-3 Account Management

Category: Common Governance, Risk, and Compliance (GRC) Requirements

Requirement

The organization manages Smart Grid information system accounts, including:

Authorizing, establishing, activating, modifying, disabling, and removing accounts;

1. Specifying account types, access rights, and privileges (e.g., individual, group, system, guest, anonymous and temporary);
2. Reviewing accounts on an organization-defined frequency; and
3. Notifying account managers when Smart Grid information system users are terminated, transferred, or Smart Grid information system usage changes.

Management approval is required prior to establishing accounts.

Supplemental Guidance

None.

Requirement Enhancements

None.

Additional Considerations

A1. The organization reviews currently active Smart Grid information system accounts on an organization-defined frequency to verify that temporary accounts and accounts of terminated or transferred users have been deactivated in accordance with organizational policy.

A2. The organization authorizes and monitors the use of guest/anonymous accounts.

A3. The organization employs automated mechanisms to support the management of Smart Grid information system accounts.

A4. The Smart Grid information system automatically terminates temporary and emergency accounts after an organization-defined time period for each type of account.

A5. The Smart Grid information system automatically disables inactive accounts after an organization-defined time period.

A6. The Smart Grid information system automatically audits account creation, modification, disabling, and termination actions and notifies, as required, appropriate individuals.

Impact Level Allocation

Low: SG.AC-3	Moderate: SG.AC-3	High: SG.AC-3

SG.AC-4 Access Enforcement

Category: Common Governance, Risk, and Compliance (GRC) Requirements

Requirement

The Smart Grid information system enforces assigned authorizations for controlling access to the Smart Grid information system in accordance with organization-defined policy.

Supplemental Guidance

None.

Requirement Enhancements

None.

Additional Considerations

A1. The organization considers the implementation of a controlled, audited, and manual override of automated mechanisms in the event of emergencies.

Impact Level Allocation

Low: SG.AC-4	Moderate: SG.AC-4	High: SG.AC-4

SG.AC-5 Information Flow Enforcement

Category: Unique Technical Requirements

Requirement

The Smart Grid information system enforces assigned authorizations for controlling the flow of information within the Smart Grid information system and between interconnected Smart Grid information systems in accordance with applicable policy.

Supplemental Guidance

Information flow control regulates where information is allowed to travel within a Smart Grid information system and between Smart Grid information systems. Specific examples of flow control enforcement can be found in boundary protection devices (e.g., proxies, gateways, guards, encrypted tunnels, firewalls, and routers) that employ rule sets or establish configuration settings that restrict Smart Grid information system services or provide a packet-filtering capability.

Requirement Enhancements

None.

Additional Considerations

A1. The Smart Grid information system enforces information flow control using explicit labels on information, source, and destination objects as a basis for flow control decisions.

A2. The Smart Grid information system enforces dynamic information flow control allowing or disallowing information flows based on changing conditions or operational considerations.

A3. The Smart Grid information system enforces information flow control using organization-defined security policy filters as a basis for flow control decisions.

A4. The Smart Grid information system enforces the use of human review for organization-defined security policy filters when the Smart Grid information system is not capable of making an information flow control decision.

A5. The Smart Grid information system provides the capability for a privileged administrator to configure, enable, and disable the organization-defined security policy filters.

Impact Level Allocation

Low: Not Selected	Moderate: Not Selected	High: Not Selected

SG.AC-6   Separation of Duties

Category: Common Technical Requirements, Integrity

Requirement

The organization—

1. Establishes and documents divisions of responsibility and separates functions as needed to eliminate conflicts of interest and to ensure independence in the responsibilities and functions of individuals/roles;

2. Enforces separation of Smart Grid information system functions through assigned access authorizations; and

3. Restricts security functions to the least amount of users necessary to ensure the security of the Smart Grid information system.

Supplemental Guidance

None.

Requirement Enhancements

None.

Additional Considerations

None.

Impact Level Allocation

| Low: Not Selected | Moderate: SG.AC-6 | High: SG.AC-6 |

SG.AC-7   Least Privilege

Category: Common Technical Requirements, Integrity

Requirement

1. The organization assigns the most restrictive set of rights and privileges or access needed by users for the performance of specified tasks; and
2. The organization configures the Smart Grid information system to enforce the most restrictive set of rights and privileges or access needed by users.

Supplemental Guidance

None.

Requirement Enhancements

None.

Additional Considerations

A1. The organization authorizes network access to organization-defined privileged commands only for compelling operational needs and documents the rationale for such access in the security plan for the Smart Grid information system.

A2. The organization authorizes access to organization-defined list of security functions (deployed in hardware, software, and firmware) and security-relevant information.

Impact Level Allocation

| Low: Not Selected | Moderate: SG.AC-7 | High: SG.AC-7 |

SG.AC-8   Unsuccessful Login Attempts

Category: Common Technical Requirements, Integrity

Requirement

The Smart Grid information system enforces a limit of organization-defined number of consecutive invalid login attempts by a user during an organization-defined time period.

Supplemental Guidance

Because of the potential for denial of service, automatic lockouts initiated by the Smart Grid information system are usually temporary and automatically released after a predetermined time period established by the organization. Permanent automatic lockouts initiated by a Smart Grid information system must be carefully considered before being used because of safety considerations and the potential for denial of service.

Requirement Enhancements

None.

Additional Considerations

A1. The Smart Grid information system automatically locks the account/node until released by an administrator when the maximum number of unsuccessful attempts is exceeded; and

A2. If a Smart Grid information system cannot perform account/node locking or delayed logins because of significant adverse impact on performance, safety, or reliability, the system employs alternative requirements or countermeasures that include the following:

  a. Real-time logging and recording of unsuccessful login attempts; and

  b. Real-time alerting of a management authority for the Smart Grid information system when the number of defined consecutive invalid access attempts is exceeded.

Impact Level Allocation

Low: SG.AC-8	Moderate: SG.AC-8	High: SG.AC-8

SG.AC-9　Smart Grid Information System Use Notification

Category: Common Technical Requirements, Integrity

Requirement

The Smart Grid information system displays an approved system use notification message or banner before granting access to the Smart Grid information system that provides privacy and security notices consistent with applicable laws, directives, policies, regulations, standards, and guidance.

Supplemental Guidance

Smart Grid information system use notification messages can be implemented in the form of warning banners displayed when individuals log in to the Smart Grid information system. Smart Grid information system use notification is intended only for Smart Grid information system access that includes an interactive interface with a human user and is not intended to call for such an interface when the interface does not currently exist.

Requirement Enhancements

None.

Additional Considerations

None.

Impact Level Allocation

| Low: SG.AC-9 | Moderate: SG.AC-9 | High: SG.AC-9 |

SG.AC-10   Previous Logon Notification

Category: Unique Technical Requirements

Requirement

The Smart Grid information system notifies the user, upon successful logon, of the date and time of the last logon and the number of unsuccessful logon attempts since the last successful logon.

Supplemental Guidance

None.

Requirement Enhancements

None.

Additional Considerations

None.

Impact Level Allocation

| Low: Not Selected | Moderate: Not Selected | High: Not Selected |

SG.AC-11   Concurrent Session Control

Category: Unique Technical Requirements

Requirement

The organization limits the number of concurrent sessions for any user on the Smart Grid information system.

Supplemental Guidance

The organization may define the maximum number of concurrent sessions for a Smart Grid information system account globally, by account type, by account, or a combination. This requirement addresses concurrent sessions for a given Smart Grid information system account and does not address concurrent sessions by a single user via multiple Smart Grid information system accounts.

Requirement Enhancements

None.

Additional Considerations

None.

Impact Level Allocation

| Low: Not Selected | Moderate: SG.AC-11 | High: SG.AC-11 |

SG.AC-12   Session Lock

Category: Unique Technical Requirements

Requirement

The Smart Grid information system—

1. Prevents further access to the Smart Grid information system by initiating a session lock after an organization-defined time period of inactivity or upon receiving a request from a user; and
2. Retains the session lock until the user reestablishes access using appropriate identification and authentication procedures.

Supplemental Guidance

A session lock is not a substitute for logging out of the Smart Grid information system.

Requirement Enhancements

None.

Additional Considerations

A1.   The Smart Grid information system session lock mechanism, when activated on a device with a display screen, places a publicly viewable pattern onto the associated display, hiding what was previously visible on the screen.

Impact Level Allocation

Low: Not Selected	Moderate: SG.AC-12	High: SG.AC-12

SG.AC-13   Remote Session Termination

Category: Unique Technical Requirements

Requirement

The Smart Grid information system terminates a remote session at the end of the session or after an organization-defined time period of inactivity.

Supplemental Guidance

None.

Requirement Enhancements

None.

Additional Considerations

A1.   Automatic session termination applies to local and remote sessions.

Impact Level Allocation

Low: Not Selected	Moderate: SG.AC-13	High: SG.AC-13

SG.AC-14   Permitted Actions without Identification or Authentication

Category: Unique Technical Requirements

Requirement

1. The organization identifies and documents specific user actions, if any, that can be performed on the Smart Grid information system without identification or authentication; and
2. Organizations identify any actions that normally require identification or authentication but may, under certain circumstances (e.g., emergencies), allow identification or authentication mechanisms to be bypassed.

Supplemental Guidance

The organization may allow limited user actions without identification and authentication (e.g., when individuals access public Web sites or other publicly accessible Smart Grid information systems.

Requirement Enhancements

1. The organization permits actions to be performed without identification and authentication only to the extent necessary to accomplish mission objectives.

Additional Considerations

None.

Impact Level Allocation

| Low: SG.AC-14 | Moderate: SG.AC-14 (1) | High: SG.AC-14 (1) |

SG.AC-15   Remote Access

Category: Unique Technical Requirements

Requirement

The organization authorizes, monitors, and manages all methods of remote access to the Smart Grid information system.

Supplemental Guidance

Remote access is any access to a Smart Grid information system by a user (or a process acting on behalf of a user) communicating through an external network (e.g., the Internet).

Requirement Enhancements

1. The organization authenticates remote access, and uses cryptography to protect the confidentiality and integrity of remote access sessions;
2. The Smart Grid information system routes all remote accesses through a limited number of managed access control points;
3. The Smart Grid information system protects wireless access to the Smart Grid information system using authentication and encryption. Note: Authentication applies to user, device, or both as necessary; and
4. The organization monitors for unauthorized remote connections to the Smart Grid information system, including scanning for unauthorized wireless access points on an

organization-defined frequency and takes appropriate action if an unauthorized connection is discovered.

Additional Considerations

A1. Remote access to Smart Grid information system component locations (e.g., control center, field locations) is enabled only when necessary, approved, authenticated, and for the duration necessary;

A2. The organization employs automated mechanisms to facilitate the monitoring and control of remote access methods;

A3. The organization authorizes remote access for privileged commands and security-relevant information only for compelling operational needs and documents the rationale for such access in the security plan for the Smart Grid information system; and

A4. The organization disables, when not intended for use, wireless networking capabilities internally embedded within Smart Grid information system components.

Impact Level Allocation

Low: SG.AC-15	Moderate: SG.AC-15 (1), (2), (3), (4)	High: SG.AC-15 (1), (2), (3), (4)

SG.AC-16   Wireless Access Restrictions

Category: Common Technical Requirements, Confidentiality

Requirement

The organization—

1. Establishes use restrictions and implementation guidance for wireless technologies; and
2. Authorizes, monitors, and manages wireless access to the Smart Grid information system.

Supplemental Guidance

None.

Requirement Enhancements

None.

Additional Considerations

A1. The organization uses authentication and encryption to protect wireless access to the Smart Grid information system; and

A2. The organization scans for unauthorized wireless access points at an organization-defined frequency and takes appropriate action if such access points are discovered.

Impact Level Allocation

Low: SG.AC-16	Moderate: SG.AC-16	High: SG.AC-16

SG.AC-17   Access Control for Portable and Mobile Devices

Category: Common Technical Requirements, Confidentiality

Requirement

The organization—

1. Establishes usage restrictions and implementation guidance for organization-controlled mobile devices, including the use of writeable, removable media and personally owned removable media;
2. Authorizes connection of mobile devices to Smart Grid information systems;
3. Monitors for unauthorized connections of mobile devices to Smart Grid information systems; and
4. Enforces requirements for the connection of mobile devices to Smart Grid information systems.

Supplemental Guidance

Specially configured mobile devices include computers with sanitized hard drives, limited applications, and additional hardening (e.g., more stringent configuration settings). Specified measures applied to mobile devices upon return from travel to locations that the organization determines to be of significant risk, include examining the device for signs of physical tampering and purging/reimaging the hard disk drive.

Requirement Enhancements

The organization—

1. Controls the use of writable, removable media in Smart Grid information systems;
2. Controls the use of personally owned, removable media in Smart Grid information systems;
3. Issues specially configured mobile devices to individuals traveling to locations that the organization determines to be of significant risk in accordance with organizational policies and procedures; and
4. Applies specified measures to mobile devices returning from locations that the organization determines to be of significant risk in accordance with organizational policies and procedures.

Additional Considerations

None.

Impact Level Allocation

Low: SG.AC-17	Moderate: SG.AC-17 (1), (2)	High: SG.AC-17 (1), (2), (3), (4)

SG.AC-18   Use of External Information Control Systems

Category: Common Governance, Risk, and Compliance (GRC) Requirements

Requirement

The organization establishes terms and conditions for authorized individuals to—

1. Access the Smart Grid information system from an external information system; and

2. Process, store, and transmit organization-controlled information using an external information system.

Supplemental Guidance

External information systems are information systems or components of information systems that are outside the authorization boundary established by the organization and for which the organization typically has no direct supervision and authority over the application of security requirements or the assessment of security requirement effectiveness.

Requirement Enhancements

1. The organization imposes restrictions on authorized individuals with regard to the use of organization-controlled removable media on external information systems.

Additional Considerations

A1. The organization prohibits authorized individuals from using an external information system to access the Smart Grid information system or to process, store, or transmit organization-controlled information except in situations where the organization (a) can verify the implementation of required security controls on the external information system as specified in the organization's security policy and security plan, or (b) has approved Smart Grid information system connection or processing agreements with the organizational entity hosting the external information system.

Impact Level Allocation

| Low: SG.AC-18 | Moderate: SG.AC-18 (1) | High: SG.AC-18 (1) |

SG.AC-19 Control System Access Restrictions

Category: Common Governance, Risk, and Compliance (GRC) Requirements

Requirement

The organization employs mechanisms in the design and implementation of a Smart Grid information system to restrict access to the Smart Grid information system from the organization's enterprise network.

Supplemental Guidance

Access to the Smart Grid information system to satisfy business requirements needs to be limited to read-only access.

Requirement Enhancements

None.

Additional Considerations

None.

Impact Level Allocation

| Low: SG.AC-19 | Moderate: SG.AC-19 | High: SG.AC-19 |

SG.AC-20 Publicly Accessible Content

Category: Common Governance, Risk, and Compliance (GRC) Requirements

Requirement

The organization—

1. Designates individuals authorized to post information onto an organizational information system that is publicly accessible;
2. Trains authorized individuals to ensure that publicly accessible information does not contain nonpublic information;
3. Reviews the proposed content of publicly accessible information for nonpublic information prior to posting onto the organizational information system;
4. Reviews the content on the publicly accessible organizational information system for nonpublic information on an organization-defined frequency; and
5. Removes nonpublic information from the publicly accessible organizational information system, if discovered.

Supplemental Guidance

Information protected under the Privacy Act and vendor proprietary information are examples of nonpublic information. This requirement addresses posting information on an organizational information system that is accessible to the general public, typically without identification or authentication.

Requirement Enhancements

None.

Additional Considerations

None.

Impact Level Allocation

| Low: SG.AC-20 | Moderate: SG.AC-20 | High: SG.AC-20 |

SG.AC-21 Passwords

Category: Common Technical Requirements, Integrity

Requirement

1. The organization develops and enforces policies and procedures for Smart Grid information system users concerning the generation and use of passwords;
2. These policies stipulate rules of complexity, based on the criticality level of the Smart Grid information system to be accessed; and
3. Passwords shall be changed regularly and are revoked after an extended period of inactivity.

Supplemental Guidance

None.

Requirement Enhancements

None.

Additional Considerations

None.

Impact Level Allocation

| Low: SG.AC-21 | Moderate: SG.AC-21 | High: SG.AC-21 |

3.8 AWARENESS AND TRAINING (SG.AT)

Smart Grid information system security awareness is a critical part of Smart Grid information system incident prevention. Implementing a Smart Grid information system security program may change the way personnel access computer programs and applications, so organizations need to design effective training programs based on individuals' roles and responsibilities.

SG.AT-1 Awareness and Training Policy and Procedures

Category: Common Governance, Risk, and Compliance (GRC) Requirements

Requirement

1. The organization develops, implements, reviews, and updates on an organization-defined frequency—

    a. A documented awareness and training security policy that addresses—

        i. The objectives, roles, and responsibilities for the awareness and training security program as it relates to protecting the organization's personnel and assets, and

        ii. The scope of the awareness and training security program as it applies to all of the organizational staff, contractors, and third parties.

    b. Procedures to address the implementation of the awareness and training security policy and associated awareness and training protection requirements.

2. Management commitment ensures compliance with the organization's security policy and other regulatory requirements; and

3. The organization ensures that the awareness and training security policy and procedures comply with applicable federal, state, local, tribal, and territorial laws and regulations.

Supplemental Guidance

The security awareness and training policy can be included as part of the general information security policy for the organization. Security awareness and training procedures can be developed for the security program in general and for a particular Smart Grid information system when required.

Requirement Enhancements

None.

Additional Considerations

None.

Impact Level Allocation

| Low: SG.AT-1 | Moderate: SG.AT-1 | High: SG.AT-1 |

SG.AT-2   Security Awareness

Category: Common Governance, Risk, and Compliance (GRC) Requirements

Requirement

The organization provides basic security awareness briefings to all Smart Grid information system users (including employees, contractors, and third parties) on an organization-defined frequency.

Supplemental Guidance

The organization determines the content of security awareness briefings based on the specific requirements of the organization and the Smart Grid information system to which personnel have authorized access.

Requirement Enhancements

None.

Additional Considerations

A1.   All Smart Grid information system design and procedure changes need to be reviewed by the organization for inclusion in the organization security awareness training; and

A2.   The organization includes practical exercises in security awareness briefings that simulate actual cyber attacks.

Impact Level Allocation

| Low: SG.AT-2 | Moderate: SG.AT-2 | High: SG.AT-2 |

SG.AT-3   Security Training

Category: Common Governance, Risk, and Compliance (GRC) Requirements

Requirement

The organization provides security-related training—

1. Before authorizing access to the Smart Grid information system or performing assigned duties;
2. When required by Smart Grid information system changes; and
3. On an organization-defined frequency thereafter.

Supplemental Guidance

The organization determines the content of security training based on assigned roles and responsibilities and the specific requirements of the organization and the Smart Grid information system to which personnel have authorized access. In addition, the organization provides Smart Grid information system managers, Smart Grid information system and network administrators, and other personnel having access to Smart Grid information system-level software, security-related training to perform their assigned duties.

Requirement Enhancements

None.

Additional Considerations

None.

Impact Level Allocation

| Low: SG.AT-3 | Moderate: SG.AT-3 | High: SG.AT-3 |

SG.AT-4   Security Awareness and Training Records

Category: Common Governance, Risk, and Compliance (GRC) Requirements

Requirement

The organization maintains a record of awareness and training for each user in accordance with the provisions of the organization's training and records retention policy.

Supplemental Guidance

None.

Requirement Enhancements

None.

Additional Considerations

None.

Impact Level Allocation

| Low: SG.AT-4 | Moderate: SG.AT-4 | High: SG.AT-4 |

SG.AT-5   Contact with Security Groups and Associations

Category: Common Governance, Risk, and Compliance (GRC) Requirements

Requirement

The organization establishes and maintains contact with security groups and associations to stay up to date with the latest recommended security practices, techniques, and technologies and to share current security-related information including threats, vulnerabilities, and incidents.

Supplemental Guidance

Security groups and associations can include special interest groups, specialized forums, professional associations, news groups, and/or peer groups of security professionals in similar organizations. The groups and associations selected are consistent with the organization's mission/business requirements.

Requirement Enhancements

None.

Additional Considerations

None.

Impact Level Allocation

| Low: Not Selected | Moderate: Not Selected | High: Not Selected |

SG.AT-6 Security Responsibility Testing

Category: Common Governance, Risk, and Compliance (GRC) Requirements

Requirement

1. The organization tests the knowledge of personnel on security policies and procedures based on their roles and responsibilities to ensure that they understand their responsibilities in securing the Smart Grid information system;
2. The organization maintains a list of security responsibilities for roles that are used to test each user in accordance with the provisions of the organization training policy; and
3. The security responsibility testing needs to be conducted on an organization-defined frequency and as warranted by technology/procedural changes.

Supplemental Guidance

None.

Requirement Enhancements

None.

Additional Considerations

None.

Impact Level Allocation

| Low: SG.AT-6 | Moderate: SG.AT-6 | High: SG.AT-6 |

SG.AT-7 Planning Process Training

Category: Common Governance, Risk, and Compliance (GRC) Requirements

Requirement

The organization includes training in the organization's planning process on the implementation of the Smart Grid information system security plans for employees, contractors, and third parties.

Supplemental Guidance

None.

Requirement Enhancements

None.

Additional Considerations

None.

Impact Level Allocation

| Low: SG.AT-7 | Moderate: SG. AT-7 | High: SG. AT-7 |

3.9 AUDIT AND ACCOUNTABILITY (SG.AU)

Periodic audits and logging of the Smart Grid information system need to be implemented to validate that the security mechanisms present during Smart Grid information system validation testing are still installed and operating correctly. These security audits review and examine a Smart Grid information system's records and activities to determine the adequacy of Smart Grid information system security requirements and to ensure compliance with established security policy and procedures. Audits also are used to detect breaches in security services through examination of Smart Grid information system logs. Logging is necessary for anomaly detection as well as forensic analysis.

SG.AU-1　Audit and Accountability Policy and Procedures

Category: Common Governance, Risk, and Compliance (GRC) Requirements

Requirement

1. The organization develops, implements, reviews, and updates on an organization-defined frequency—
    a. A documented audit and accountability security policy that addresses—
        i. The objectives, roles, and responsibilities for the audit and accountability security program as it relates to protecting the organization's personnel and assets; and
        ii. The scope of the audit and accountability security program as it applies to all of the organizational staff, contractors, and third parties.
    b. Procedures to address the implementation of the audit and accountability security policy and associated audit and accountability protection requirements.
2. Management commitment ensures compliance with the organization's security policy and other regulatory requirements; and
3. The organization ensures that the audit and accountability security policy and procedures comply with applicable federal, state, local, tribal, and territorial laws and regulations.

Supplemental Guidance

The audit and accountability policy can be included as part of the general security policy for the organization. Procedures can be developed for the security program in general and for a particular Smart Grid information system when required.

Requirement Enhancements

None.

Additional Considerations

None.

Impact Level Allocation

| Low: SG.AU-1 | Moderate: SG.AU-1 | High: SG.AU-1 |

SG.AU-2   Auditable Events

Category: Common Technical Requirements, Integrity

Requirement

The organization—

1. Develops, based on a risk assessment, the Smart Grid information system list of auditable events on an organization-defined frequency;
2. Includes execution of privileged functions in the list of events to be audited by the Smart Grid information system; and
3. Revises the list of auditable events based on current threat data, assessment of risk, and post-incident analysis.

Supplemental Guidance

The purpose of this requirement is for the organization to identify events that need to be auditable as significant and relevant to the security of the Smart Grid information system.

Requirement Enhancements

1. The organization should audit activities associated with configuration changes to the Smart Grid information system.

Additional Considerations

None.

Impact Level Allocation

| Low: SG.AU-2 | Moderate: SG.AU-2 (1) | High: SG.AU-2 (1) |

SG.AU-3   Content of Audit Records

Category: Common Technical Requirements, Integrity

Requirement

The Smart Grid information system produces audit records for each event. The record contains the following information:

- Data and time of the event,
- The component of the Smart Grid information system where the event occurred,
- Type of event,
- User/subject identity, and
- The outcome of the events.

Supplemental Guidance

None.

Requirement Enhancements

None.

Additional Considerations

A1. The Smart Grid information system provides the capability to include additional, more detailed information in the audit records for audit events identified by type, location, or subject; and

A2. The Smart Grid information system provides the capability to centrally manage the content of audit records generated by individual components throughout the Smart Grid information system.

Impact Level Allocation

Low: SG.AU-3	Moderate: SG.AU-3	High: SG.AU-3

SG.AU-4 Audit Storage Capacity

Category: Common Technical Requirements, Integrity

Requirement

The organization allocates organization-defined audit record storage capacity and configures auditing to reduce the likelihood of such capacity being exceeded.

Supplemental Guidance

The organization considers the types of auditing to be performed and the audit processing requirements when allocating audit storage capacity.

Requirement Enhancements

None.

Additional Considerations

None.

Impact Level Allocation

Low: SG.AU-4	Moderate: SG.AU-4	High: SG.AU-4

SG.AU-5 Response to Audit Processing Failures

Category: Common Governance, Risk, and Compliance (GRC) Requirements

Requirement

The Smart Grid information system—

1. Alerts designated organizational officials in the event of an audit processing failure; and
2. Executes an organization-defined set of actions to be taken (e.g., shutdown Smart Grid information system, overwrite oldest audit records, and stop generating audit records).

Supplemental Guidance

Audit processing failures include software/hardware errors, failures in the audit capturing mechanisms, and audit storage capacity being reached or exceeded.

Requirement Enhancements

1. The Smart Grid information system provides a warning when allocated audit record storage volume reaches an organization-defined percentage of maximum audit record storage capacity; and
2. The Smart Grid information system provides a real-time alert for organization defined audit failure events.

Additional Considerations

None.

Impact Level Allocation

| Low: SG.AU-5 | Moderate: SG.AU-5 | High: SG.AU-5 (1), (2) |

SG.AU-6   Audit Monitoring, Analysis, and Reporting

Category: Common Governance, Risk, and Compliance (GRC) Requirements

Requirement

The organization—

1. Reviews and analyzes Smart Grid information system audit records on an organization-defined frequency for indications of inappropriate or unusual activity and reports findings to management authority; and
2. Adjusts the level of audit review, analysis, and reporting within the Smart Grid information system when a change in risk occurs to organizational operations, organizational assets, or individuals.

Supplemental Guidance

Organizations increase the level of audit monitoring and analysis activity within the Smart Grid information system based on, for example, law enforcement information, intelligence information, or other credible sources of information.

Requirement Enhancements

None.

Additional Considerations

A1.   The Smart Grid information system employs automated mechanisms to integrate audit review, analysis, and reporting into organizational processes for investigation and response to suspicious activities;

A2.   The organization analyzes and correlates audit records across different repositories to gain organization-wide situational awareness;

A3.   The Smart Grid information system employs automated mechanisms to centralize audit review and analysis of audit records from multiple components within the Smart Grid information system; and

A4. The organization integrates analysis of audit records with analysis of performance and network monitoring information to further enhance the ability to identify inappropriate or unusual activity.

Impact Level Allocation

| Low: SG.AU-6 | Moderate: SG.AU-6 | High: SG.AU-6 |

SG.AU-7 Audit Reduction and Report Generation

Category: Common Governance, Risk, and Compliance (GRC) Requirements

Requirement

The Smart Grid information system provides an audit reduction and report generation capability.

Supplemental Guidance

Audit reduction and reporting may support near real-time analysis and after-the-fact investigations of security incidents.

Requirement Enhancements

None.

Additional Considerations

A1. The Smart Grid information system provides the capability to automatically process audit records for events of interest based on selectable event criteria

Impact Level Allocation

| Low: Not Selected | Moderate: SG.AU-7 | High: SG.AU-7 |

SG.AU-8 Time Stamps

Category: Common Governance, Risk, and Compliance (GRC) Requirements

Requirement

The Smart Grid information system uses internal system clocks to generate time stamps for audit records.

Supplemental Guidance

Time stamps generated by the information system include both date and time, as defined by the organization.

Requirement Enhancements

1. The Smart Grid information system synchronizes internal Smart Grid information system clocks on an organization-defined frequency using an organization-defined time source.

Additional Considerations

None.

Impact Level Allocation

| Low: SG.AU-8 | Moderate: SG.AU-8 (1) | High: SG.AU-8 (1) |

SG.AU-9 Protection of Audit Information

Category: Common Governance, Risk, and Compliance (GRC) Requirements

Requirement

The Smart Grid information system protects audit information and audit tools from unauthorized access, modification, and deletion.

Supplemental Guidance

Audit information includes, for example, audit records, audit settings, and audit reports.

Requirement Enhancements

None.

Additional Considerations

A1. The Smart Grid information system produces audit records on hardware-enforced, write-once media.

Impact Level Allocation

Low: SG.AU-9	Moderate: SG.AU-9	High: SG.AU-9

SG.AU-10 Audit Record Retention

Category: Common Governance, Risk, and Compliance (GRC) Requirements

Requirement

The organization retains audit logs for an organization-defined time period to provide support for after-the-fact investigations of security incidents and to meet regulatory and organizational information retention requirements.

Supplemental Guidance

None.

Requirement Enhancements

None.

Additional Considerations

None.

Impact Level Allocation

Low: SG.AU-10	Moderate: SG.AU-10	High: SG.AU-10

SG.AU-11 Conduct and Frequency of Audits

Category: Common Governance, Risk, and Compliance (GRC) Requirements

Requirement

The organization conducts audits on an organization-defined frequency to assess conformance to specified security requirements and applicable laws and regulations.

Supplemental Guidance

Audits can be either in the form of internal self-assessment (sometimes called first-party audits) or independent, third-party audits.

Requirement Enhancements

None.

Additional Considerations

None.

Impact Level Allocation

| Low: SG.AU-11 | Moderate: SG.AU-11 | High: SG.AU-11 |

SG.AU-12 Auditor Qualification

Category: Common Governance, Risk, and Compliance (GRC) Requirements

Requirement

The organization's audit program specifies auditor qualifications.

Supplemental Guidance

Security auditors need to—

1. Understand the Smart Grid information system and the associated operating practices;
2. Understand the risk involved with the audit; and
3. Understand the organization cyber security and the Smart Grid information system policy and procedures.

Requirement Enhancements

None.

Additional Considerations

A1. The organization assigns auditor and Smart Grid information system administration functions to separate personnel.

Impact Level Allocation

| Low: SG.AU-12 | Moderate: SG.AU-12 | High: SG.AU-12 |

SG.AU-13 Audit Tools

Category: Common Governance, Risk, and Compliance (GRC) Requirements

Requirement

The organization specifies the rules and conditions of use of audit tools.

Supplemental Guidance

Access to Smart Grid information systems audit tools needs to be protected to prevent any possible misuse or compromise.

Requirement Enhancements

None.

Additional Considerations

None.

Impact Level Allocation

| Low: SG.AU-13 | Moderate: SG.AU-13 | High: SG.AU-13 |

SG.AU-14 Security Policy Compliance

Category: Common Governance, Risk, and Compliance (GRC) Requirements

Requirement

The organization demonstrates compliance to the organization's security policy through audits in accordance with the organization's audit program.

Supplemental Guidance

Periodic audits of the Smart Grid information system are implemented to demonstrate compliance to the organization's security policy. These audits—

1. Assess whether the defined cyber security policies and procedures, including those to identify security incidents, are being implemented and followed;
2. Document and ensure compliance to organization policies and procedures;
3. Identify security concerns, validate that the Smart Grid information system is free from security compromises, and provide information on the nature and extent of compromises should they occur;
4. Validate change management procedures and ensure that they produce an audit trail of reviews and approvals of all changes;
5. Verify that security mechanisms and management practices present during Smart Grid information system validation are still in place and functioning;
6. Ensure reliability and availability of the Smart Grid information system to support safe operation; and
7. Continuously improve performance.

Requirement Enhancements

None.

Additional Considerations

None.

Impact Level Allocation

| Low: SG.AU-14 | Moderate: SG.AU-14 | High: SG.AU-14 |

SG.AU-15   Audit Generation

Category: Common Technical Requirements, Integrity

Requirement

The Smart Grid information system—

1. Provides audit record generation capability and generates audit records for the selected list of auditable events; and
2. Provides audit record generation capability and allows authorized users to select auditable events at the organization-defined Smart Grid information system components.

Supplemental Guidance

Audit records can be generated from various components within the Smart Grid information system.

Requirement Enhancements

None.

Additional Considerations

A1. The Smart Grid information system provides the capability to compile audit records from multiple components within the Smart Grid information system into a Smart Grid information system-wide audit trail that is time-correlated to within an organization-defined level of tolerance for relationship between time stamps of individual records in the audit trail.

Impact Level Allocation

Low: SG.AU-15	Moderate: SG.AU-15	High: SG.AU-15

SG.AU-16   Non-Repudiation

Category: Unique Technical Requirements

Requirement

The Smart Grid information system protects against an individual falsely denying having performed a particular action.

Supplemental Guidance

Non-repudiation protects individuals against later claims by an author of not having authored a particular document, a sender of not having transmitted a message, a receiver of not having received a message, or a signatory of not having signed a document. Non-repudiation services are implemented using various techniques (e.g., digital signatures, digital message receipts, and logging).

Requirement Enhancements

None.

Additional Considerations

None.

Impact Level Allocation

| Low: Not Selected | Moderate: Not Selected | High: SG.AU-16 |

3.10 SECURITY ASSESSMENT AND AUTHORIZATION (SG.CA)

Security assessments include monitoring and reviewing the performance of Smart Grid information system. Internal checking methods, such as compliance audits and incident investigations, allow the organization to determine the effectiveness of the security program. Finally, through continuous monitoring, the organization regularly reviews compliance of the Smart Grid information systems. If deviations or nonconformance exist, it may be necessary to revisit the original assumptions and implement appropriate corrective actions.

SG.CA-1 Security Assessment and Authorization Policy and Procedures

Category: Common Governance, Risk, and Compliance (GRC) Requirements

Requirement

1. The organization develops, implements, reviews, and updates on an organization-defined frequency—
   a. A documented security assessment and authorization policy that addresses—
      i. The objectives, roles, and responsibilities for the security assessment and authorization security program as it relates to protecting the organization's personnel and assets; and
      ii. The scope of the security assessment and authorization security program as it applies to all of the organizational staff and third-party contractors; and
   b. Procedures to address the implementation of the security assessment and authorization policy and associated security assessment and authorization protection requirements;
2. Management commitment ensures compliance with the organization's security assessment and authorization security policy and other regulatory requirements; and
3. The organization ensures that the security assessment and authorization security policy and procedures comply with applicable federal, state, local, tribal, and territorial laws and regulations.

Supplemental Guidance

The authorization to operate and security assessment policies can be included as part of the general information security policy for the organization. Authorization to operate and security assessment procedures can be developed for the security program in general and for a particular Smart Grid information system when required. The organization defines significant change to a Smart Grid information system for security reauthorizations.

Requirement Enhancements

None.

Additional Considerations

None.

Impact Level Allocation

| Low: SG.CA-1 | Moderate: SG.CA-1 | High: SG.CA-1 |

SG.CA-2   Security Assessments

Category: Common Governance, Risk, and Compliance (GRC) Requirements

Requirement

The organization—

1. Develops a security assessment plan that describes the scope of the assessment including—

    a. Security requirements and requirement enhancements under assessment;

    b. Assessment procedures to be used to determine security requirement effectiveness; and

    c. Assessment environment, assessment team, and assessment roles and responsibilities;

2. Assesses the security requirements in the Smart Grid information system on an organization-defined frequency to determine the extent the requirements are implemented correctly, operating as intended, and producing the desired outcome with respect to meeting the security requirements for the Smart Grid information system;

3. Produces a security assessment report that documents the results of the assessment; and

4. Provides the results of the security requirements assessment to a management authority.

Supplemental Guidance

The organization assesses the security requirements in a Smart Grid information system as part of authorization or reauthorization to operate and continuous monitoring. Previous security assessment results may be reused to the extent that they are still valid and are supplemented with additional assessments as needed.

Requirement Enhancements

None.

Additional Considerations

A1.   The organization employs an independent assessor or assessment team to conduct an assessment of the security requirements in the Smart Grid information system.

Impact Level Allocation

| Low: SG.CA-2 | Moderate: SG.CA-2 | High: SG.CA-2 |

SG.CA-3   Continuous Improvement

Category: Common Governance, Risk, and Compliance (GRC) Requirements

Requirement

The organization's security program implements continuous improvement practices to ensure that industry lessons learned and best practices are incorporated into Smart Grid information system security policies and procedures.

Supplemental Guidance

None.

Requirement Enhancements

None.

Additional Considerations

None.

Impact Level Allocation

| Low: Not Selected | Moderate: Not Selected | High: Not Selected |

SG.CA-4   Smart Grid Information System Connections

Category: Common Governance, Risk, and Compliance (GRC) Requirements

Requirement

The organization—

1. Authorizes all connections from the Smart Grid information system to other information systems;
2. Documents the Smart Grid information system connections and associated security requirements for each connection; and
3. Monitors the Smart Grid information system connections on an ongoing basis, verifying enforcement of documented security requirements.

Supplemental Guidance

The organization considers the risk that may be introduced when a Smart Grid information system is connected to other information systems, both internal and external to the organization, with different security requirements. Risk considerations also include Smart Grid information systems sharing the same networks.

Requirement Enhancements

None.

Additional Considerations

A1.   All external Smart Grid information system and communication connections are identified and protected from tampering or damage.

Impact Level Allocation

| Low: SG.CA-4 | Moderate: SG.CA-4 | High: SG.CA-4 |

SG.CA-5 Security Authorization to Operate

Category: Common Governance, Risk, and Compliance (GRC) Requirements

Requirement

1. The organization authorizes the Smart Grid information system for processing before operation and updates the authorization based on an organization-defined frequency or when a significant change occurs to the Smart Grid information system; and
2. A management authority signs and approves the security authorization to operate. Security assessments conducted in support of security authorizations need to be reviewed on an organization-defined frequency.

Supplemental Guidance

The organization assesses the security mechanisms implemented within the Smart Grid information system prior to security authorization to operate.

Requirement Enhancements

None.

Additional Considerations

None.

Impact Level Allocation

Low: SG.CA-5	Moderate: SG.CA-5	High: SG.CA-5

SG.CA-6 Continuous Monitoring

Category: Common Governance, Risk, and Compliance (GRC) Requirements

Requirement

The organization establishes a continuous monitoring strategy and implements a continuous monitoring program that includes:

1. Ongoing security requirements assessments in accordance with the organizational continuous monitoring strategy; and
2. Reporting the security state of the Smart Grid information system to management authority on an organization-defined frequency.

Supplemental Guidance

A continuous monitoring program allows an organization to maintain the security authorization to operate of a Smart Grid information system over time in a dynamic operational environment with changing threats, vulnerabilities, technologies, and missions/business processes.

The selection of an appropriate subset of security requirements for continuous monitoring is based on the impact level of the Smart Grid information system, the specific security requirements selected by the organization, and the level of assurance that the organization requires.

Requirement Enhancements

None.

Additional Considerations

A1. The organization employs an independent assessor or assessment team to monitor the security requirements in the Smart Grid information system on an ongoing basis;

A2. The organization includes as part of security requirements continuous monitoring, periodic, unannounced, in-depth monitoring, penetration testing, and red team exercises; and

A3. The organization uses automated support tools for continuous monitoring.

Impact Level Allocation

| Low: SG.CA-6 | Moderate: SG.CA-6 | High: SG.CA-6 |

3.11 CONFIGURATION MANAGEMENT (SG.CM)

The organization's security program needs to implement policies and procedures that create a process by which the organization manages and documents all configuration changes to the Smart Grid information system. A comprehensive change management process needs to be implemented and used to ensure that only approved and tested changes are made to the Smart Grid information system configuration. Smart Grid information systems need to be configured properly to maintain optimal operation. Therefore, only tested and approved changes should be allowed on a Smart Grid information system. Vendor updates and patches need to be thoroughly tested on a non-production Smart Grid information system setup before being introduced into the production environment to ensure that no adverse effects occur.

SG.CM-1 Configuration Management Policy and Procedures

Category: Common Governance, Risk, and Compliance (GRC) Requirements

Requirement

1. The organization develops, implements, reviews, and updates on an organization-defined frequency—

    a. A documented configuration management security policy that addresses—

        i. The objectives, roles, and responsibilities for the configuration management security program as it relates to protecting the organization's personnel and assets; and

        ii. The scope of the configuration management security program as it applies to all of the organizational staff, contractors, and third parties; and

    b. Procedures to address the implementation of the configuration management security policy and associated configuration management protection requirements;

2. Management commitment ensures compliance with the organization's security policy and other regulatory requirements; and

3. The organization ensures that the configuration management security policy and procedures comply with applicable federal, state, local, tribal, and territorial laws and regulations.

Supplemental Guidance

The configuration management policy can be included as part of the general system security policy for the organization. Configuration management procedures can be developed for the security program in general and for a particular Smart Grid information system when required.

Requirement Enhancements

None.

Additional Considerations

None.

Impact Level Allocation

| Low: SG.CM-1 | Moderate: SG.CM-1 | High: SG.CM-1 |

SG.CM-2   Baseline Configuration

Category: Common Governance, Risk, and Compliance (GRC) Requirements

Requirement

The organization develops, documents, and maintains a current baseline configuration of the Smart Grid information system and an inventory of the Smart Grid information system's constituent components. The organization reviews and updates the baseline configuration as an integral part of Smart Grid information system component installations.

Supplemental Guidance

Maintaining the baseline configuration involves updating the baseline as the Smart Grid information system changes over time and keeping previous baselines for possible rollback.

Requirement Enhancements

None.

Additional Considerations

A1. The organization maintains a baseline configuration for development and test environments that is managed separately from the operational baseline configuration; and

A2. The organization employs automated mechanisms to maintain an up-to-date, complete, accurate, and readily available baseline configuration of the Smart Grid information system.

Impact Level Allocation

| Low: SG.CM-2 | Moderate: SG.CM-2 | High: SG.CM-2 |

SG.CM-3   Configuration Change Control

Category: Common Governance, Risk, and Compliance (GRC) Requirements

Requirement

The organization—

1. Authorizes and documents changes to the Smart Grid information system;
2. Retains and reviews records of configuration-managed changes to the Smart Grid information system;
3. Audits activities associated with configuration-managed changes to the Smart Grid information system; and
4. Tests, validates, and documents configuration changes (e.g., patches and updates) before installing them on the operational Smart Grid information system.

Supplemental Guidance

Configuration change control includes changes to the configuration settings for the Smart Grid information system and those IT products (e.g., operating systems, firewalls, routers) that are components of the Smart Grid information system. The organization includes emergency changes in the configuration change control process, including changes resulting from the remediation of flaws.

Requirement Enhancements

None.

Additional Considerations

None.

Impact Level Allocation

| Low: Not Selected | Moderate: SG.CM-3 | High: SG.CM-3 |

SG.CM-4   Monitoring Configuration Changes

Category: Common Governance, Risk, and Compliance (GRC) Requirements

Requirement

1. The organization implements a process to monitor changes to the Smart Grid information system;
2. Prior to change implementation and as part of the change approval process, the organization analyzes changes to the Smart Grid information system for potential security impacts; and
3. After the Smart Grid information system is changed, the organization checks the security features to ensure that the features are still functioning properly.

Supplemental Guidance

Security impact analysis may also include an assessment of risk to understand the impact of the changes and to determine if additional safeguards and countermeasures are required. The organization considers Smart Grid information system safety and security interdependencies.

Requirement Enhancements

None.

Additional Considerations

None.

Impact Level Allocation

| Low: SG.CM-4 | Moderate: SG.CM-4 | High: SG.CM-4 |

SG.CM-5   Access Restrictions for Configuration Change

Category: Common Governance, Risk, and Compliance (GRC) Requirements

Requirement

The organization—

1. Defines, documents, and approves individual access privileges and enforces access restrictions associated with configuration changes to the Smart Grid information system;
2. Generates, retains, and reviews records reflecting all such changes;
3. Establishes terms and conditions for installing any hardware, firmware, or software on Smart Grid information system devices; and
4. Conducts audits of Smart Grid information system changes at an organization-defined frequency and if/when suspected unauthorized changes have occurred.

Supplemental Guidance

Planned or unplanned changes to the hardware, software, and/or firmware components of the Smart Grid information system may affect the overall security of the Smart Grid information system. Only authorized individuals should be allowed to obtain access to Smart Grid information system components for purposes of initiating changes, including upgrades, and modifications. Maintaining records is important for supporting after-the-fact actions should the organization become aware of an unauthorized change to the Smart Grid information system.

Requirement Enhancements

None.

Additional Considerations

A1.   The organization employs automated mechanisms to enforce access restrictions and support auditing of the enforcement actions.

Impact Level Allocation

| Low: Not Selected | Moderate: SG.CM-5 | High: SG.CM-5 |

SG.CM-6   Configuration Settings

Category: Common Governance, Risk, and Compliance (GRC) Requirements

Requirement

The organization—

1. Establishes configuration settings for components within the Smart Grid information system;
2. Monitors and controls changes to the configuration settings in accordance with organizational policies and procedures;
3. Documents changed configuration settings;
4. Identifies, documents, and approves exceptions from the configuration settings; and
5. Enforces the configuration settings in all components of the Smart Grid information system.

Supplemental Guidance

None.

Requirement Enhancements

None.

Additional Considerations

A1. The organization employs automated mechanisms to centrally manage, apply, and verify configuration settings;

A2. The organization employs automated mechanisms to respond to unauthorized changes to configuration settings; and

A3. The organization incorporates detection of unauthorized, security-relevant configuration changes into the organization's incident response capability to ensure that such detected events are tracked, monitored, corrected, and available for historical purposes.

Impact Level Allocation

Low: SG.CM-6	Moderate: SG.CM-6	High: SG.CM-6

SG.CM-7   Configuration for Least Functionality

Category: Common Technical Requirements, Integrity

Requirement

1. The organization configures the Smart Grid information system to provide only essential capabilities and specifically prohibits and/or restricts the use of functions, ports, protocols, and/or services as defined in an organizationally generated "prohibited and/or restricted" list; and
2. The organization reviews the Smart Grid information system on an organization-defined frequency or as deemed necessary to identify and restrict unnecessary functions, ports, protocols, and/or services.

Supplemental Guidance

The organization considers disabling unused or unnecessary physical and logical ports on Smart Grid information system components to prevent unauthorized connection of devices, and considers designing the overall system to enforce a policy of least functionality.

Requirement Enhancements

None.

Additional Considerations

None.

Impact Level Allocation

| Low: SG.CM-7 | Moderate: SG.CM-7 | High: SG.CM-7 |

SG.CM-8   Component Inventory

Category: Common Technical Requirements, Integrity

Requirement

The organization develops, documents, and maintains an inventory of the components of the Smart Grid information system that—

1. Accurately reflects the current Smart Grid information system configuration;
2. Provides the proper level of granularity deemed necessary for tracking and reporting and for effective property accountability;
3. Identifies the roles responsible for component inventory;
4. Updates the inventory of system components as an integral part of component installations, system updates, and removals; and
5. Ensures that the location (logical and physical) of each component is included within the Smart Grid information system boundary.

Supplemental Guidance

The organization determines the appropriate level of granularity for any Smart Grid information system component included in the inventory that is subject to management control (e.g., tracking, reporting).

Requirement Enhancements

None.

Additional Considerations

A1.   The organization updates the inventory of the information system components as an integral part of component installations and information system updates;

A2.   The organization employs automated mechanisms to maintain an up-to-date, complete, accurate, and readily available inventory of information system components; and

A3.   The organization employs automated mechanisms to detect the addition of unauthorized components or devise into the environment and disables access by components or devices or notifies designated officials.

Impact Level Allocation

| Low: SG.CM-8 | Moderate: SG.CM-8 | High: SG.CM-8 |

SG.CM-9 Addition, Removal, and Disposal of Equipment

Category: Common Governance, Risk, and Compliance (GRC) Requirements

Requirement

1. The organization implements policy and procedures to address the addition, removal, and disposal of all Smart Grid information system equipment; and
2. All Smart Grid information system components and information are documented, identified, and tracked so that their location and function are known.

Supplemental Guidance

The policies and procedures should consider the sensitivity of critical security parameters such as passwords, cryptographic keys, and personally identifiable information such as name and social security numbers.

Requirement Enhancements

None.

Additional Considerations

None.

Impact Level Allocation

Low: SG.CM-9	Moderate: SG.CM-9	High: SG.CM-9

SG.CM-10 Factory Default Settings Management

Category: Common Governance, Risk, and Compliance (GRC) Requirements

Requirement

1. The organization policy and procedures require the management of all factory default settings (e.g., authentication credentials, user names, configuration settings, and configuration parameters) on Smart Grid information system components and applications; and
2. The factory default settings should be changed upon installation and if used during maintenance.

Supplemental Guidance

Many Smart Grid information system devices and software are shipped with factory default settings to allow for initial installation and configuration.

Requirement Enhancements

None.

Additional Considerations

A1. The organization replaces default usernames whenever possible; and

A2. Default passwords of applications, operating systems, database management systems, or other programs must be changed within an organizational-defined time period.

Impact Level Allocation

| Low: SG.CM-10 | Moderate: SG.CM-10 | High: SG.CM-10 |

SG.CM-11   Configuration Management Plan

Category: Common Governance, Risk, and Compliance (GRC) Requirements

Requirement

The organization develops and implements a configuration management plan for the Smart Grid information system that—

1. Addresses roles, responsibilities, and configuration management processes and procedures;
2. Defines the configuration items for the Smart Grid information system;
3. Defines when (in the system development life cycle) the configuration items are placed under configuration management;
4. Defines the means for uniquely identifying configuration items throughout the system development life cycle; and
5. Defines the process for managing the configuration of the controlled items.

Supplemental Guidance

The configuration management plan defines processes and procedures for how configuration management is used to support system development life cycle activities.

Requirement Enhancements

None.

Additional Considerations

None.

Impact Level Allocation

| Low: SG.CM-11 | Moderate: SG.CM-11 | High: SG.CM-11 |

3.12   CONTINUITY OF OPERATIONS (SG.CP)

Continuity of operations addresses the capability to continue or resume operations of a Smart Grid information system in the event of disruption of normal system operation. The ability for the Smart Grid information system to function after an event is dependent on implementing continuity of operations policies, procedures, training, and resources. The security requirements recommended under the continuity of operations family provide policies and procedures for roles and responsibilities, training, testing, plan updates, alternate storage sites, alternate command and control methods, alternate control centers, recovery and reconstitution and fail-safe response.

SG.CP-1   Continuity of Operations Policy and Procedures

Category: Common Governance, Risk, and Compliance (GRC) Requirements

Requirement

1. The organization develops, implements, reviews, and updates on an organization-defined frequency—
   a. A documented continuity of operations security policy that addresses—
      i. The objectives, roles, and responsibilities for the continuity of operations security program as it relates to protecting the organization's personnel and assets; and
      ii. The scope of the continuity of operations security program as it applies to all of the organizational staff, contractors, and third parties; and
   b. Procedures to address the implementation of the continuity of operations security policy and associated continuity of operations protection requirements;
2. Management commitment ensures compliance with the organization's security policy and other regulatory requirements; and
3. The organization ensures that the continuity of operations security policy and procedures comply with applicable federal, state, local, tribal, and territorial laws and regulations.

Supplemental Guidance

The continuity of operations policy can be included as part of the general information security policy for the organization. Continuity of operations procedures can be developed for the security program in general, and for a particular Smart Grid information system, when required.

Requirement Enhancements

None.

Additional Considerations

None.

Impact Level Allocation

| Low: SG.CP-1 | Moderate: SG.CP-1 | High: SG.CP-1 |

SG.CP-2   Continuity of Operations Plan

Category: Common Governance, Risk, and Compliance (GRC) Requirements

Requirement

1. The organization develops and implements a continuity of operations plan dealing with the overall issue of maintaining or reestablishing operations in case of an undesirable interruption for a Smart Grid information system;
2. The plan addresses roles, responsibilities, assigned individuals with contact information, and activities associated with restoring Smart Grid information system operations after a disruption or failure; and
3. A management authority reviews and approves the continuity of operations plan.

Supplemental Guidance

A continuity of operations plan addresses both business continuity planning and recovery of Smart Grid information system operations. Development of a continuity of operations plan is a process to identify procedures for safe Smart Grid information system operation while recovering from a Smart Grid information system disruption. The plan requires documentation of critical Smart Grid information system functions that need to be recovered.

Requirement Enhancements

None.

Additional Considerations

A1. The organization performs a root cause analysis for the event and submits any findings from the analysis to management.

Impact Level Allocation

| Low: SG.CP-2 | Moderate: SG.CP-2 | High: SG.CP-2 |

SG.CP-3 Continuity of Operations Roles and Responsibilities

Category: Common Governance, Risk, and Compliance (GRC) Requirements

Requirement

The continuity of operations plan—

1. Defines the roles and responsibilities of the various employees and contractors in the event of a significant incident; and
2. Identifies responsible personnel to lead the recovery and response effort if an incident occurs.

Supplemental Guidance

None.

Requirement Enhancements

None.

Additional Considerations

None.

Impact Level Allocation

| Low: SG.CP-3 | Moderate: SG.CP-3 | High: SG.CP-3 |

SG.CP-4 Continuity of Operations Training

Category: Common Governance, Risk, and Compliance (GRC) Requirements

Requirement

The organization trains personnel in their continuity of operations roles and responsibilities with respect to the Smart Grid information system and provides refresher training on an organization-defined frequency.

Supplemental Guidance

None.

Requirement Enhancements

None.

Additional Considerations

None.

Impact Level Allocation

| Low: SG.CP-4 | Moderate: SG.CP-4 | High: SG.CP-4 |

SG.CP-5   Continuity of Operations Plan Testing

Category: Common Governance, Risk, and Compliance (GRC) Requirements

Requirement

1. The continuity of operations plan is tested to determine its effectiveness and results are documented;
2. A management authority reviews the documented test results and initiates corrective actions, if necessary; and
3. The organization tests the continuity of operations plan for the Smart Grid information system on an organization-defined frequency, using defined tests.

Supplemental Guidance

None.

Requirement Enhancements

1. The organization coordinates continuity of operations plan testing and exercises with all affected organizational elements.

Additional Considerations

A1. The organization employs automated mechanisms to test/exercise the continuity of operations plan; and

A2. The organization tests/exercises the continuity of operations plan at the alternate processing site to familiarize Smart Grid information system operations personnel with the facility and available resources and to evaluate the site's capabilities to support continuity of operations.

Impact Level Allocation

| Low: SG.CP-5 | Moderate: SG. CP-5 (1) | High: SG. CP-5 (1) |

SG.CP-6   Continuity of Operations Plan Update

Category: Common Governance, Risk, and Compliance (GRC) Requirements

Requirement

The organization reviews the continuity of operations plan for the Smart Grid information system and updates the plan to address Smart Grid information system, organizational, and technology changes or problems encountered during plan implementation, execution, or testing on an organization-defined frequency.

Supplemental Guidance

Organizational changes include changes in mission, functions, or business processes supported by the Smart Grid information system. The organization communicates the changes to appropriate organizational elements.

Requirement Enhancements

None.

Additional Considerations

None.

Impact Level Allocation

| Low: SG.CP-6 | Moderate: SG.CP-6 | High: SG.CP-6 |

SG.CP-7   Alternate Storage Sites

Category: Common Governance, Risk, and Compliance (GRC) Requirements

Requirement

The organization determines the requirement for an alternate storage site and initiates any necessary agreements.

Supplemental Guidance

The Smart Grid information system backups and the transfer rate of backup information to the alternate storage site are performed on an organization-defined frequency.

Requirement Enhancements

1. The organization identifies potential accessibility problems at the alternative storage site in the event of an area-wide disruption or disaster and outlines explicit mitigation actions;
2. The organization identifies an alternate storage site that is geographically separated from the primary storage site so it is not susceptible to the same hazards; and
3. The organization configures the alternate storage site to facilitate timely and effective recovery operations.

Additional Considerations

None.

Impact Level Allocation

| Low: Not Selected | Moderate: SG.CP-7 (1), (2) | High: SG.SG.CP-7 (1), (2), (3) |

SG.CP-8   Alternate Telecommunication Services

Category: Common Governance, Risk, and Compliance (GRC) Requirements

Requirement

The organization identifies alternate telecommunication services for the Smart Grid information system and initiates necessary agreements to permit the resumption of operations for the safe operation of the Smart Grid information system within an organization-defined time period when the primary Smart Grid information system capabilities are unavailable.

Supplemental Guidance

Alternate telecommunication services required to resume operations within the organization-defined time period are either available at alternate organization sites or contracts with vendors need to be in place to support alternate telecommunication services for the Smart Grid information system.

Requirement Enhancements

1. Primary and alternate telecommunication service agreements contain priority-of-service provisions in accordance with the organization's availability requirements;
2. Alternate telecommunication services do not share a single point of failure with primary telecommunication services;
3. Alternate telecommunication service providers need to be sufficiently separated from primary service providers so they are not susceptible to the same hazards; and
4. Primary and alternate telecommunication service providers need to have adequate contingency plans.

Additional Considerations

None.

Impact Level Allocation

Low: Not Selected	Moderate: SG.CP-8 (1), (4)	High: SG. CP-8 (1), (2), (3), (4)

SG.CP-9   Alternate Control Center

Category: Common Governance, Risk, and Compliance (GRC) Requirements

Requirement

The organization identifies an alternate control center, necessary telecommunications, and initiates any necessary agreements to permit the resumption of Smart Grid information system operations for critical functions within an organization-prescribed time period when the primary control center is unavailable.

Supplemental Guidance

Equipment, telecommunications, and supplies required to resume operations within the organization-prescribed time period need to be available at the alternative control center or by a contract in place to support delivery to the site.

Requirement Enhancements

1. The organization identifies an alternate control center that is geographically separated from the primary control center so it is not susceptible to the same hazards;
2. The organization identifies potential accessibility problems to the alternate control center in the event of an area-wide disruption or disaster and outlines explicit mitigation actions; and
3. The organization develops alternate control center agreements that contain priority-of-service provisions in accordance with the organization's availability requirements.

Additional Considerations

A1. The organization fully configures the alternate control center and telecommunications so that they are ready to be used as the operational site supporting a minimum required operational capability; and

A2. The organization ensures that the alternate processing site provides information security measures equivalent to that of the primary site.

Impact Level Allocation

Low: Not Selected	Moderate: SG.CP-9 (1), (2), (3)	High: SG.CP-9 (1), (2), (3)

SG.CP-10 Smart Grid Information System Recovery and Reconstitution

Category: Common Governance, Risk, and Compliance (GRC) Requirements

Requirement

The organization provides the capability to recover and reconstitute the Smart Grid information system to a known secure state after a disruption, compromise, or failure.

Supplemental Guidance

Smart Grid information system recovery and reconstitution to a known secure state means that—

1. All Smart Grid information system parameters (either default or organization-established) are set to secure values;
2. Security-critical patches are reinstalled;
3. Security-related configuration settings are reestablished;
4. Smart Grid information system documentation and operating procedures are available;
5. Application and Smart Grid information system software is reinstalled and configured with secure settings;
6. Information from the most recent, known secure backups is loaded; and
7. The Smart Grid information system is fully tested.

Requirement Enhancements

1. The organization provides compensating security controls (including procedures or mechanisms) for the organization-defined circumstances that inhibit recovery to a known, secure state; and
2. The organization provides the capability to reimage Smart Grid information system components in accordance with organization-defined restoration time periods from configuration-controlled and integrity-protected media images representing a secure, operational state for the components.

Additional Considerations

None.

Impact Level Allocation

| Low: SG.CP-10 | Moderate: SG.CP-10 (1) | High: SG.CP-10 (1), (2) |

SG.CP-11   Fail-Safe Response

Category: Common Governance, Risk, and Compliance (GRC) Requirements

Requirement

The Smart Grid information system has the ability to execute an appropriate fail-safe procedure upon the loss of communications with other Smart Grid information systems or the loss of the Smart Grid information system itself.

Supplemental Guidance

In the event of a loss of communication between the Smart Grid information system and the operational facilities, the on-site instrumentation needs to be capable of executing a procedure that provides the maximum protection to the controlled infrastructure. For the electric sector, this may be to alert the operator of the failure and then do nothing (i.e., let the electric grid continue to operate). The organization defines what "loss of communications" means (e.g., 5 seconds or 5 minutes without communications). The organization then defines the appropriate fail-safe process for its industry.

Requirement Enhancements

None.

Additional Considerations

A1.   The Smart Grid information system preserves the organization-defined state information in failure.

Impact Level Allocation

| Low: Not Selected | Moderate: Not Selected | High: SG.CP-11 |

3.13 IDENTIFICATION AND AUTHENTICATION (SG.IA)

Identification and authentication is the process of verifying the identity of a user, process, or device, as a prerequisite for granting access to resources in a Smart Grid information system.

SG.IA-1   Identification and Authentication Policy and Procedures

Category: Common Governance, Risk, and Compliance (GRC) Requirements

Requirement

1. The organization develops, implements, reviews, and updates on an organization-defined frequency—
   a. A documented identification and authentication security policy that addresses—
      i. The objectives, roles, and responsibilities for the identification and authentication security program as it relates to protecting the organization's personnel and assets; and
      ii. The scope of the identification and authentication security program as it applies to all of the organizational staff, contractors, and third parties; and
   b. Procedures to address the implementation of the identification and authentication security policy and associated identification and authentication protection requirements;
2. Management commitment ensures compliance with the organization's security policy and other regulatory requirements; and
3. The organization ensures that the identification and authentication security policy and procedures comply with applicable federal, state, local, tribal, and territorial laws and regulations.

Supplemental Guidance

The identification and authentication policy can be included as part of the general security policy for the organization. Identification and authentication procedures can be developed for the security program in general and for a particular Smart Grid information system when required.

Requirement Enhancements

None.

Additional Considerations

None.

Impact Level Allocation

Low: SG.IA-1	Moderate: SG.IA-1	High: SG.IA-1

SG.IA-2   Identifier Management

Category: Common Governance, Risk, and Compliance (GRC) Requirements

Requirement

The organization receives authorization from a management authority to assign a user or device identifier.

Supplemental Guidance

None.

Requirement Enhancements

None.

Additional Considerations

A1. The organization archives previous user or device identifiers; and

A2. The organization selects an identifier that uniquely identifies an individual or device.

Impact Level Allocation

| Low: SG.IA-2 | Moderate: SG.IA-2 | High: SG.IA-2 |

SG.IA-3 Authenticator Management

Category: Common Governance, Risk, and Compliance (GRC) Requirements

Requirement

The organization manages Smart Grid information system authentication credentials for users and devices by—

1. Defining initial authentication credential content, such as defining password length and composition, tokens;
2. Establishing administrative procedures for initial authentication credential distribution; lost, compromised, or damaged authentication credentials; and revoking authentication credentials;
3. Changing/refreshing authentication credentials on an organization-defined frequency; and
4. Specifying measures to safeguard authentication credentials.

Supplemental Guidance

Measures to safeguard user authentication credentials include maintaining possession of individual authentication credentials, not loaning or sharing authentication credentials with others, and reporting lost or compromised authentication credentials immediately.

Requirement Enhancements

None.

Additional Considerations

A1. The organization employs automated tools to determine if authentication credentials are sufficiently strong to resist attacks intended to discover or otherwise compromise the authentication credentials; and

A2. The organization requires unique authentication credentials be provided by vendors and manufacturers of Smart Grid information system components.

Impact Level Allocation

| Low: SG.IA-3 | Moderate: SG.IA-3 | High: SG.IA-3 |

SG.IA-4　User Identification and Authentication

Category: Unique Technical Requirements

Requirement

The Smart Grid information system uniquely identifies and authenticates users (or processes acting on behalf of users).

Supplemental Guidance

None.

Requirement Enhancements

None.

Additional Considerations

A1.　The Smart Grid information system uses multifactor authentication for—

    a.　Remote access to non-privileged accounts;

    b.　Local access to privileged accounts; and

    c.　Remote access to privileged accounts.

Impact Level Allocation

Low: SG.IA-4	Moderate: SG.IA-4	High: SG.IA-4

SG.IA-5　Device Identification and Authentication

Category: Unique Technical Requirements

Requirement

The Smart Grid information system uniquely identifies and authenticates an organization-defined list of devices before establishing a connection.

Supplemental Guidance

The devices requiring unique identification and authentication may be defined by type, by specific device, or by a combination of type and device as deemed appropriate by the organization.

Requirement Enhancements

1. The Smart Grid information system authenticates devices before establishing remote network connections using bidirectional authentication between devices that is cryptographically based; and

2. The Smart Grid information system authenticates devices before establishing network connections using bidirectional authentication between devices that is cryptographically based.

Additional Considerations

None.

Impact Level Allocation

| Low: Not Selected | Moderate: SG.IA-5 (1), (2) | High: SG.IA-5 (1), (2) |

SG.IA-6 Authenticator Feedback

Category: Unique Technical Requirements

Requirement

The authentication mechanisms in the Smart Grid information system obscure feedback of authentication information during the authentication process to protect the information from possible exploitation/use by unauthorized individuals.

Supplemental Guidance

The Smart Grid information system obscures feedback of authentication information during the authentication process (e.g., displaying asterisks when a user types in a password). The feedback from the Smart Grid information system does not provide information that would allow an unauthorized user to compromise the authentication mechanism.

Requirement Enhancements

None.

Additional Considerations

None.

Impact Level Allocation

| Low: SG.IA-6 | Moderate: SG.IA-6 | High: SG.IA-6 |

3.14 INFORMATION AND DOCUMENT MANAGEMENT (SG.ID)

Information and document management is generally a part of the organization records retention and document management system. Digital and hardcopy information associated with the development and execution of a Smart Grid information system is important and sensitive, and need to be managed. Smart Grid information system design, operations data and procedures, risk analyses, business impact studies, risk tolerance profiles, etc., contain sensitive organization information and need to be protected. This information must be protected and verified that the appropriate versions are retained.

The following are the requirements for Information and Document Management that need to be supported and implemented by the organization to protect the Smart Grid information system.

SG.ID-1 Information and Document Management Policy and Procedures

Category: Common Governance, Risk, and Compliance (GRC) Requirements

Requirement

1. The organization develops, implements, reviews, and updates on an organization-defined frequency—

    a. A Smart Grid information and document management policy that addresses—

i. The objectives, roles and responsibilities for the information and document management security program as it relates to protecting the organization's personnel and assets;

ii. The scope of the information and document management security program as it applies to all the organizational staff, contractors, and third parties;

iii. The retrieval of written and electronic records, equipment, and other media for the Smart Grid information system; and

iv. The destruction of written and electronic records, equipment, and other media for the Smart Grid information system; and

b. Procedures to address the implementation of the information and document management security policy and associated Smart Grid information system information and document management protection requirements;

2. Management commitment ensures compliance of the organization's security policy and other regulatory requirements; and

3. The organization ensures that the Smart Grid information system information and document management policy and procedures comply with applicable federal, state, local, tribal, and territorial laws and regulations.

Supplemental Guidance

The information and document management policy may be included as part of the general information security policy for the organization. The information and document management procedures can be developed for the security program in general and for a particular Smart Grid information system when required. The organization employs appropriate measures to ensure that long-term records and information can be retrieved (e.g., converting the data to a newer format, retaining older equipment that can read the data). Destruction includes the method of disposal such as shredding of paper records, erasing of disks or other electronic media, or physical destruction.

Requirement Enhancements

None.

Additional Considerations

None.

Impact Level Allocation

| Low: SG.ID-1 | Moderate: SG.ID-1 | High: SG.ID-1 |

SG.ID-2   Information and Document Retention

Category: Common Governance, Risk, and Compliance (GRC) Requirements

Requirement

1. The organization develops policies and procedures detailing the retention of organization information;

2. The organization performs legal reviews of the retention policies to ensure compliance with all applicable laws and regulations;
3. The organization manages Smart Grid information system-related data including establishing retention policies and procedures for both electronic and paper data; and
4. The organization manages access to Smart Grid information system-related data based on assigned roles and responsibilities.

Supplemental Guidance

The retention procedures address retention/destruction issues for all applicable information media.

Requirement Enhancements

None.

Additional Considerations

None.

Impact Level Allocation

| Low: SG.ID-2 | Moderate: SG.ID-2 | High: SG.ID-2 |

SG.ID-3   Information Handling

Category: Common Governance, Risk, and Compliance (GRC) Requirements

Requirement

Organization-implemented policies and procedures detailing the handling of information are developed and reviewed on an organization-defined frequency.

Supplemental Guidance

Written policies and procedures detail access, sharing, copying, transmittal, distribution, and disposal or destruction of Smart Grid information system information. These policies or procedures include the periodic review of all information to ensure that it is properly handled.

Requirement Enhancements

None.

Additional Considerations

None.

Impact Level Allocation

| Low: SG.ID-3 | Moderate: SG.ID-3 | High: SG.ID-3 |

SG.ID-4   Information Exchange

Category: Common Governance, Risk, and Compliance (GRC) Requirements

Requirement

Agreements are established for the exchange of information, firmware, and software between the organization and external parties such as third parties, vendors and contractors.

Supplemental Guidance

None.

Requirement Enhancements

None.

Additional Considerations

A1. If a specific device needs to communicate with another device outside the Smart Grid information system, communications need to be limited to only the devices that need to communicate.

Impact Level Allocation

| Low: SG.ID-4 | Moderate: SG.ID-4 | High: SG.ID-4 |

SG.ID-5   Automated Labeling

Category: Common Governance, Risk, and Compliance (GRC) Requirements

Requirement

The Smart Grid information system automatically labels information in storage, in process, and in transmission in accordance with—

1. Access control requirements;
2. Special dissemination, handling, or distribution instructions; and
3. Otherwise as required by the Smart Grid information system security policy.

Supplemental Guidance

Automated labeling refers to labels employed on internal data structures (e.g., records, buffers, files) within the Smart Grid information system. Such labels are often used to implement access control and flow control policies.

Requirement Enhancements

None.

Additional Considerations

A1. The Smart Grid information system maintains the binding of the label to the information.

Impact Level Allocation

| Low: Not Selected | Moderate: Not Selected | High: Not Selected |

3.15 INCIDENT RESPONSE (SG.IR)

Incident response addresses the capability to continue or resume operations of a Smart Grid information system in the event of disruption of normal Smart Grid information system operation. Incident response entails the preparation, testing, and maintenance of specific policies and procedures to enable the organization to recover the Smart Grid information system's operational status after the occurrence of a disruption. Disruptions can come from natural disasters, such as earthquakes, tornados, floods, or from manmade events like riots, terrorism, or vandalism. The ability for the Smart Grid information system to function after such an event is directly dependent on implementing policies, procedures, training, and resources in place ahead of time using the organization's planning process. The security requirements recommended under the incident response family provide policies and procedures for incident response monitoring, handling, reporting, testing, training, recovery, and reconstitution of the Smart Grid information systems for an organization.

SG.IR-1 Incident Response Policy and Procedures

Category: Common Governance, Risk, and Compliance (GRC) Requirements

Requirement

1. The organization develops, implements, reviews, and updates on an organization-defined frequency—
    a. A documented incident response security policy that addresses—
        i. The objectives, roles, and responsibilities for the incident response security program as it relates to protecting the organization's personnel and assets; and
        ii. The scope of the incident response security program as it applies to all of the organizational staff, contractors, and third parties; and
    b. Procedures to address the implementation of the incident response security policy and associated incident response protection requirements;
2. Management commitment ensures compliance with the organization's security policy and other regulatory requirements;
3. The organization ensures that the incident response security policy and procedures comply with applicable federal, state, local, tribal, and territorial laws and regulations; and
4. The organization identifies potential interruptions and classifies them as to "cause," "effects," and "likelihood."

Supplemental Guidance

The incident response policy can be included as part of the general information security policy for the organization. Incident response procedures can be developed for the security program in general, and for a particular Smart Grid information system, when required. The various types of incidents that may result from system intrusion need to be identified and classified as to their effects and likelihood so that a proper response can be formulated for each potential incident. The organization determines the impact to each Smart Grid system and the consequences associated with loss of one or more of the Smart Grid information systems.

Requirement Enhancements

None.

Additional Considerations

None.

Impact Level Allocation

| Low: SG.IR-1 | Moderate: SG.IR-1 | High: SG.IR-1 |

SG.IR-2   Incident Response Roles and Responsibilities

Category: Common Governance, Risk, and Compliance (GRC) Requirements

Requirement

1. The organization's Smart Grid information system security plan defines the specific roles and responsibilities in relation to various types of incidents; and
2. The plan identifies responsible personnel to lead the response effort if an incident occurs. Response teams need to be formed, including Smart Grid information system and other process owners, to reestablish operations.

Supplemental Guidance

The organization's Smart Grid information system security plan defines the roles and responsibilities of the various employees, contractors, and third parties in the event of an incident. The response teams have a major role in the interruption identification and planning process.

Requirement Enhancements

None.

Additional Considerations

None.

Impact Level Allocation

| Low: SG.IR-2 | Moderate: SG.IR-2 | High: SG.IR-2 |

SG.IR-3   Incident Response Training

Category: Common Governance, Risk, and Compliance (GRC) Requirements

Requirement

Personnel are trained in their incident response roles and responsibilities with respect to the Smart Grid information system and receive refresher training on an organization-defined frequency.

Supplemental Guidance

None.

Requirement Enhancements

None.

Additional Considerations

A1. The organization incorporates Smart Grid information system simulated events into continuity of operations training to facilitate effective response by personnel in crisis situations; and

A2. The organization employs automated mechanisms to provide a realistic Smart Grid information system training environment.

Impact Level Allocation

Low: SG.IR-3	Moderate: SG.IR-3	High: SG.IR-3

SG.IR-4 Incident Response Testing and Exercises

Category: Common Governance, Risk, and Compliance (GRC) Requirements

Requirement

The organization tests and/or exercises the incident response capability for the information system at an organization-defined frequency using organization-defined tests and/or exercises to determine the incident response effectiveness and documents the results.

Supplemental Guidance

None.

Requirement Enhancements

None.

Additional Considerations

The organization employs automated mechanisms to more thoroughly and effectively test/exercise the incident response capability

Impact Level Allocation

Low: SG.IR-4	Moderate: SG.IR-4	High: SG.IR-4

SG.IR-5 Incident Handling

Category: Common Governance, Risk, and Compliance (GRC) Requirements

Requirement

The organization—

1. Implements an incident handling capability for security incidents that includes preparation, detection and analysis, containment, mitigation, and recovery;
2. Integrates incident handling procedures with continuity of operations procedures; and
3. Incorporates lessons learned from incident handling activities into incident response procedures.

Supplemental Guidance

None.

Requirement Enhancements

None.

Additional Considerations

A1. The organization employs automated mechanisms to administer and support the incident handling process.

Impact Level Allocation

| Low: SG.IR-5 | Moderate: SG.IR-5 | High: SG.IR-5 |

SG.IR-6   Incident Monitoring

Category: Common Governance, Risk, and Compliance (GRC) Requirements

Requirement

The organization tracks and documents Smart Grid information system and network security incidents.

Supplemental Guidance

None.

Requirement Enhancements

None.

Additional Considerations

A1. The organization employs automated mechanisms to assist in the tracking of security incidents and in the collection and analysis of incident information.

Impact Level Allocation

| Low: SG.IR-6 | Moderate: SG.IR-6 | High: SG.IR-6 |

SG.IR-7   Incident Reporting

Category: Common Governance, Risk, and Compliance (GRC) Requirements

Requirement

1. The organization incident reporting procedure includes:

    a. What is a reportable incident;

    b. The granularity of the information reported;

    c. Who receives the report; and

    d. The process for transmitting the incident information.

2. Detailed incident data is reported in a manner that complies with applicable federal, state, local, tribal, and territorial laws and regulations.

Supplemental Guidance

None.

Requirement Enhancements

None.

Additional Considerations

A1. The organization employs automated mechanisms to assist in the reporting of security incidents.

Impact Level Allocation

| Low: SG.IR-7 | Moderate: SG.IR-7 | High: SG.IR-7 |

SG.IR-8 Incident Response Investigation and Analysis

Category: Common Governance, Risk, and Compliance (GRC) Requirements

Requirement

1. The organization policies and procedures include an incident response investigation and analysis program;
2. The organization includes investigation and analysis of Smart Grid information system incidents in the planning process; and
3. The organization develops, tests, deploys, and documents an incident investigation and analysis process.

Supplemental Guidance

The organization documents its policies and procedures to show that investigation and analysis of incidents are included in the planning process. The procedures ensure that the Smart Grid information system is capable of providing event data to the proper personnel for analysis and for developing mitigation steps.

Requirement Enhancements

None.

Additional Considerations

None.

Impact Level Allocation

| Low: SG.IR-8 | Moderate: SG.IR-8 | High: SG.IR-8 |

SG.IR-9 Corrective Action

Category: Common Governance, Risk, and Compliance (GRC) Requirements

Requirement

1. The organization reviews investigation results and determines corrective actions needed; and
2. The organization includes processes and mechanisms in the planning to ensure that corrective actions identified as the result of cyber security and Smart Grid information system incidents are fully implemented.

Supplemental Guidance

The organization encourages and promotes cross-industry incident information exchange and cooperation to learn from the experiences of others.

Requirement Enhancements

None.

Additional Considerations

None.

Impact Level Allocation

| Low: SG.IR-9 | Moderate: SG.IR-9 | High: SG.IR-9 |

SG.IR-10   Smart Grid Information System Backup

Category: Common Governance, Risk, and Compliance (GRC) Requirements

Requirement

The organization—

1. Conducts backups of user-level information contained in the Smart Grid information system on an organization-defined frequency;
2. Conducts backups of Smart Grid information system-level information (including Smart Grid information system state information) contained in the Smart Grid information system on an organization-defined frequency;
3. Conducts backups of information system documentation including security-related documentation on an organization-defined frequency consistent with recovery time; and
4. Protects the confidentiality and integrity of backup information at the storage location.

Supplemental Guidance

The protection of Smart Grid information system backup information while in transit is beyond the scope of this requirement.

Requirement Enhancements

1. The organization tests backup information at an organization-defined frequency to verify media reliability and information integrity;
2. The organization selectively uses backup information in the restoration of Smart Grid information system functions as part of continuity of operations testing; and
3. The organization stores backup copies of the operating system and other critical Smart Grid information system software in a separate facility or in a fire-rated container that is not collocated with the operational software.

Additional Considerations

None.

Impact Level Allocation

| Low: SG.IR-10 | Moderate: SG.IR-10 (1) | High: SG.IR-10 (1), (2), (3) |

SG.IR-11 Coordination of Emergency Response

Category: Common Governance, Risk, and Compliance (GRC) Requirements

Requirement

The organization's security policies and procedures delineate how the organization implements its emergency response plan and coordinates efforts with law enforcement agencies, regulators, Internet service providers and other relevant organizations in the event of a security incident.

Supplemental Guidance

The organization expands relationships with local emergency response personnel to include information sharing and coordinated response to cyber security incidents.

Requirement Enhancements

None.

Additional Considerations

None.

Impact Level Allocation

Low: SG.IR-11	Moderate: SG.IR-11	High: SG.IR-11

3.16 SMART GRID INFORMATION SYSTEM DEVELOPMENT AND MAINTENANCE (SG.MA)

Security is most effective when it is designed into the Smart Grid information system and sustained, through effective maintenance, throughout the life cycle of the Smart Grid information system. Maintenance activities encompass appropriate policies and procedures for performing routine and preventive maintenance on the components of a Smart Grid information system. This includes the use of both local and remote maintenance tools and management of maintenance personnel.

SG.MA-1 Smart Grid Information System Maintenance Policy and Procedures

Category: Common Governance, Risk, and Compliance (GRC) Requirements

Requirement

1. The organization develops, implements, reviews, and updates on an organization-defined frequency—

    a. A documented Smart Grid information system maintenance security policy that addresses—

        i. The objectives, roles, and responsibilities for the Smart Grid information system maintenance security program as it relates to protecting the organization's personnel and assets; and

        ii. The scope of the Smart Grid information system maintenance security program as it applies to all of the organizational staff, contractors, and third parties; and

b. Procedures to address the implementation of the Smart Grid information system maintenance security policy and associated Smart Grid information system maintenance protection requirements;

2. Management commitment ensures compliance with the organization's security policy and other regulatory requirements; and

3. The organization ensures that the Smart Grid information system maintenance security policy and procedures comply with applicable federal, state, local, tribal, and territorial laws and regulations.

Supplemental Guidance

The Smart Grid information system maintenance policy can be included as part of the general information security policy for the organization. Smart Grid information system maintenance procedures can be developed for the security program in general and for a particular Smart Grid information system when required.

Requirement Enhancements

None.

Additional Considerations

None.

Impact Level Allocation

| Low: SG.MA-1 | Moderate: SG.MA-1 | High: SG.MA-1 |

SG.MA-2 Legacy Smart Grid Information System Upgrades

Category: Common Governance, Risk, and Compliance (GRC) Requirements

Requirement

The organization develops policies and procedures to upgrade existing legacy Smart Grid information systems to include security mitigating measures commensurate with the organization's risk tolerance and the risk to the Smart Grid information system.

Supplemental Guidance

None.

Requirement Enhancements

None.

Additional Considerations

None.

Impact Level Allocation

| Low: SG.MA-2 | Moderate: SG.MA-2 | High: SG.MA-2 |

SG.MA-3 Smart Grid Information System Maintenance

Category: Common Governance, Risk, and Compliance (GRC) Requirements

Requirement

The organization—

1. Schedules, performs, documents, and reviews records of maintenance and repairs on Smart Grid information system components in accordance with manufacturer or vendor specifications and/or organizational requirements;
2. Explicitly approves the removal of the Smart Grid information system or Smart Grid information system components from organizational facilities for off-site maintenance or repairs;
3. Sanitizes the equipment to remove all critical/sensitive information from associated media prior to removal from organizational facilities for off-site maintenance or repairs;
4. Checks all potentially impacted security requirements to verify that the requirements are still functioning properly following maintenance or repair actions; and
5. Makes and secures backups of critical Smart Grid information system software, applications, and data for use if the operating system becomes corrupted or destroyed.

Supplemental Guidance

All maintenance activities to include routine, scheduled maintenance and repairs, and unplanned maintenance are controlled whether performed on site or remotely and whether the equipment is serviced on site or removed to another location. Maintenance procedures that require the physical removal of any Smart Grid information system component needs to be documented, listing the date, time, reason for removal, estimated date of reinstallation, and name personnel removing components.

Requirement Enhancements

1. The organization maintains maintenance records for the Smart Grid information system that include:
    a. The date and time of maintenance;
    b. Name of the individual performing the maintenance;
    c. Name of escort, if necessary;
    d. A description of the maintenance performed; and
    e. A list of equipment removed or replaced (including identification numbers, if applicable).

Additional Considerations

A1. The organization employs automated mechanisms to schedule and document maintenance and repairs as required, producing up-to-date, accurate, complete, and available records of all maintenance and repair actions needed, in process, and completed.

Impact Level Allocation

| Low: SG.MA-3 | Moderate: SG.MA-3 | High: SG.MA-3 (1) |

SG.MA-4 Maintenance Tools

Category: Common Governance, Risk, and Compliance (GRC) Requirements

Requirement

The organization approves and monitors the use of Smart Grid information system maintenance tools.

Supplemental Guidance

The requirement addresses security-related issues when the hardware, firmware, and software are brought into the Smart Grid information system for diagnostic and repair actions.

Requirement Enhancements

None.

Additional Considerations

A1. The organization requires approval from a management authority explicitly authorizing removal of equipment from the facility;

A2. The organization inspects all maintenance tools carried into a facility by maintenance personnel for obvious improper modifications;

A3. The organization checks all media containing diagnostic and test programs for malicious code before the media are used in the Smart Grid information system; and

A4. The organization employs automated mechanisms to restrict the use of maintenance tools to authorized personnel only.

Impact Level Allocation

Low: SG.MA-4	Moderate: SG.MA-4	High: SG.MA-4

SG.MA-5 Maintenance Personnel

Category: Common Governance, Risk, and Compliance (GRC) Requirements

Requirement

1. The organization documents authorization and approval policies and procedures for maintaining a list of personnel authorized to perform maintenance on the Smart Grid information system; and

2. When maintenance personnel do not have needed access authorizations, organizational personnel with appropriate access authorizations supervise maintenance personnel during the performance of maintenance activities on the Smart Grid information system.

Supplemental Guidance

Maintenance personnel need to have appropriate access authorization to the Smart Grid information system when maintenance activities allow access to organizational information that could result in a future compromise of availability, integrity, or confidentiality.

Requirement Enhancements

None.

Additional Considerations

None.

Impact Level Allocation

| Low: SG.MA-5 | Moderate: SG.MA-5 | High: SG.MA-5 |

SG.MA-6   Remote Maintenance

Category: Common Governance, Risk, and Compliance (GRC) Requirements

Requirement

The organization policy and procedures for remote maintenance include:

1. Authorization and monitoring the use of remote maintenance and diagnostic activities;
2. Use of remote maintenance and diagnostic tools;
3. Maintenance records for remote maintenance and diagnostic activities;
4. Termination of all remote maintenance sessions; and
5. Management of authorization credentials used during remote maintenance.

Supplemental Guidance

None.

Requirement Enhancements

The organization—

1. Requires that remote maintenance or diagnostic services be performed from an information system that implements a level of security at least as high as that implemented on the Smart Grid information system being serviced; or
2. Removes the component to be serviced from the Smart Grid information system and prior to remote maintenance or diagnostic services, sanitizes the component (with regard to organizational information) before removal from organizational facilities and after the service is performed, sanitizes the component (with regard to potentially malicious software) before returning the component to the Smart Grid information system.

Additional Considerations

A1.   The organization requires that remote maintenance sessions are protected through the use of a strong authentication credential; and

A2.   The organization requires that (a) maintenance personnel notify the Smart Grid information system administrator when remote maintenance is planned (e.g., date/time), and (b) a management authority approves the remote maintenance.

Impact Level Allocation

| Low: SG.MA-6 | Moderate: SG.MA-6 | High: SG.MA-6 (1) |

SG.MA-7 Timely Maintenance

Category: Common Governance, Risk, and Compliance (GRC) Requirements

Requirement

The organization obtains maintenance support and spare parts for an organization-defined list of security-critical Smart Grid information system components.

Supplemental Guidance

The organization specifies those Smart Grid information system components that, when not operational, result in increased risk to organizations or individuals because the security functionality intended by that component is not being provided.

Requirement Enhancements

None.

Additional Considerations

None.

Impact Level Allocation

Low: SG.MA-7	Moderate: SG.MA-7	High: SG.MA-7

3.17 MEDIA PROTECTION (SG.MP)

The security requirements under the media protection family provide policy and procedures for limiting access to media to authorized users. Security measures also exist for distribution and handling requirements as well as storage, transport, sanitization (removal of information from digital media), destruction, and disposal of the media. Media assets include compact discs; digital video discs; erasable, programmable read-only memory; tapes; printed reports; and documents.

SG.MP-1 Media Protection Policy and Procedures

Category: Common Governance, Risk, and Compliance (GRC) Requirements

Requirement

1. The organization develops, implements, reviews, and updates on an organization-defined frequency—

    a. A documented media protection security policy that addresses—

        i. The objectives, roles, and responsibilities for the media protection security program as it relates to protecting the organization's personnel and assets; and

        ii. The scope of the media protection security program as it applies to all of the organizational staff, contractors, and third parties; and

    b. Procedures to address the implementation of the media protection security policy and associated media protection requirements;

2. Management commitment ensures compliance with the organization's security policy and other regulatory requirements; and

3. The organization ensures that the media protection security policy and procedures comply with applicable federal, state, local, tribal, and territorial laws and regulations.

Supplemental Guidance

The media protection policy can be included as part of the general security policy for the organization. Media protection procedures can be developed for the security program in general and for a particular Smart Grid information system when required.

Requirement Enhancements

None.

Additional Considerations

None.

Impact Level Allocation

| Low: SG.MP-1 | Moderate: SG.MP-1 | High: SG.MP-1 |

SG.MP-2   Media Sensitivity Level

Category: Common Governance, Risk, and Compliance (GRC) Requirements

Requirement

The sensitivity level of media indicates the protection required commensurate with the impact of compromise.

Supplemental Guidance

These media sensitivity levels provide guidance for access and control to include sharing, copying, transmittal, and distribution appropriate for the level of protection required.

Requirement Enhancements

None.

Additional Considerations

None.

Impact Level Allocation

| Low: SG.MP-2 | Moderate: SG.MP-2 | High: SG.MP-2 |

SG.MP-3   Media Marking

Category: Common Governance, Risk, and Compliance (GRC) Requirements

Requirement

The organization marks removable Smart Grid information system media and Smart Grid information system output in accordance with organization-defined policy and procedures.

Supplemental Guidance

Smart Grid information system markings refer to the markings employed on external media (e.g., video displays, hardcopy documents output from the Smart Grid information system).

External markings are distinguished from internal markings (i.e., the labels used on internal data structures within the Smart Grid information system).

Requirement Enhancements

None.

Additional Considerations

None.

Impact Level Allocation

| Low: Not Selected | Moderate: SG.MP-3 | High: SG.MP-3 |

SG.MP-4 Media Storage

Category: Common Governance, Risk, and Compliance (GRC) Requirements

Requirement

The organization physically manages and stores Smart Grid information system media within protected areas. The sensitivity of the material determines how the media are stored.

Supplemental Guidance

None.

Requirement Enhancements

None.

Additional Considerations

None.

Impact Level Allocation

| Low: SG.MP-4 | Moderate: SG.MP-4 | High: SG.MP-4 |

SG.MP-5 Media Transport

Category: Common Governance, Risk, and Compliance (GRC) Requirements

Requirement

The organization—

1. Protects organization-defined types of media during transport outside controlled areas using organization-defined security measures;
2. Maintains accountability for Smart Grid information system media during transport outside controlled areas; and
3. Restricts the activities associated with transport of such media to authorized personnel.

Supplemental Guidance

A controlled area is any space for which the organization has confidence that the physical and procedural protections provided are sufficient to meet the requirements established for protecting the information and Smart Grid information system.

Requirement Enhancements

None.

Additional Considerations

A1. The organization employs an identified custodian throughout the transport of Smart Grid information system media; and

A2. The organization documents activities associated with the transport of Smart Grid information system media using an organization-defined Smart Grid information system of records.

Impact Level Allocation

| Low: SG.MP-5 | Moderate: SG.MP-5 | High: SG.MP-5 |

SG.MP-6   Media Sanitization and Disposal

Category: Common Governance, Risk, and Compliance (GRC) Requirements

Requirement

The organization sanitizes Smart Grid information system media before disposal or release for reuse. The organization tests sanitization equipment and procedures to verify correct performance on an organization-defined frequency.

Supplemental Guidance

Sanitization is the process of removing information from media such that data recovery is not possible.

Requirement Enhancements

The organization tracks, documents, and verifies media sanitization and disposal actions.

Additional Considerations

None.

Impact Level Allocation

| Low: SG.MP-6 | Moderate: SG.MP-6 (1) | High: SG.MP-6 (1) |

3.18 PHYSICAL AND ENVIRONMENTAL SECURITY (SG.PE)

Physical and environmental security encompasses protection of physical assets from damage, misuse, or theft. Physical access control, physical boundaries, and surveillance are examples of security practices used to ensure that only authorized personnel are allowed to access Smart Grid information systems and components. Environmental security addresses the safety of assets from damage from environmental concerns. Physical and environmental security addresses protection from environmental threats.

SG.PE-1   Physical and Environmental Security Policy and Procedures

Category: Common Governance, Risk, and Compliance (GRC) Requirements

Requirement

1. The organization develops, implements, reviews, and updates on an organization-defined frequency—
   a. A documented physical and environmental security policy that addresses—
      i. The objectives, roles, and responsibilities for the physical and environmental security program as it relates to protecting the organization's personnel and assets; and
      ii. The scope of the physical and environmental security program as it applies to all of the organizational staff, contractors, and third parties; and
   b. Procedures to address the implementation of the physical and environmental security policy and associated physical and environmental protection requirements;
2. Management commitment ensures compliance with the organization's security policy and other regulatory requirements; and
3. The organization ensures that the physical and environmental security policy and procedures comply with applicable federal, state, local, tribal, and territorial laws and regulations.

Supplemental Guidance

The organization may include the physical and environmental security policy as part of the general security policy for the organization.

Requirement Enhancements

None.

Additional Considerations

None.

Impact Level Allocation

| Low: SG.PE-1 | Moderate: SG.PE-1 | High: SG.PE-1 |

SG.PE-2   Physical Access Authorizations

Category: Common Governance, Risk, and Compliance (GRC) Requirements

Requirement

1. The organization develops and maintains lists of personnel with authorized access to facilities containing Smart Grid information systems and issues appropriate authorization credentials (e.g., badges, identification cards); and
2. Designated officials within the organization review and approve access lists on an organization-defined frequency, removing from the access lists personnel no longer requiring access.

Supplemental Guidance

None.

Requirement Enhancements

None.

Additional Considerations

A1. The organization authorizes physical access to the facility where the Smart Grid information system resides based on position or role;

A2. The organization requires multiple forms of identification to gain access to the facility where the Smart Grid information system resides; and

A3. The organization requires multifactor authentication to gain access to the facility where the Smart Grid information system resides.

Impact Level Allocation

Low: SG.PE-2	Moderate: SG.PE-2	High: SG.PE-2

SG.PE-3 Physical Access

Category: Common Governance, Risk, and Compliance (GRC) Requirements

Requirement

The organization—

1. Enforces physical access authorizations for all physical access points to the facility where the Smart Grid information system resides;
2. Verifies individual access authorizations before granting access to the facility;
3. Controls entry to facilities containing Smart Grid information systems;
4. Secures keys, combinations, and other physical access devices;
5. Inventories physical access devices on a periodic basis; and
6. Changes combinations, keys, and authorization credentials on an organization-defined frequency and when keys are lost, combinations are compromised, individual credentials are lost, or individuals are transferred or terminated.

Supplemental Guidance

Physical access devices include keys, locks, combinations, and card readers. Workstations and associated peripherals connected to (and part of) an organizational Smart Grid information system may be located in areas designated as publicly accessible with access to such devices being safeguarded.

Requirement Enhancements

1. The organization requires physical access mechanisms to Smart Grid information system assets in addition to physical access mechanisms to the facility; and
2. The organization employs hardware to deter unauthorized physical access to Smart Grid information system devices.

Additional Considerations

A1. The organization ensures that every physical access point to the facility where the Smart Grid information system resides is guarded or alarmed and monitored on an organization-defined frequency.

Impact Level Allocation

| Low: SG.PE-3 | Moderate: SG.PE-3 (2) | High: SG.PE-3 (1), (2) |

SG.PE-4 Monitoring Physical Access

Category: Common Governance, Risk, and Compliance (GRC) Requirements

Requirement

The organization—

1. Monitors physical access to the Smart Grid information system to detect and respond to physical security incidents;
2. Reviews physical access logs on an organization-defined frequency;
3. Coordinates results of reviews and investigations with the organization's incident response capability; and
4. Ensures that investigation of and response to detected physical security incidents, including apparent security violations or suspicious physical access activities, are part of the organization's incident response capability.

Supplemental Guidance

None.

Requirement Enhancements

None.

Additional Considerations

A1. The organization installs and monitors real-time physical intrusion alarms and surveillance equipment; and

A2. The organization implements automated mechanisms to recognize potential intrusions and initiates designated response actions.

Impact Level Allocation

| Low: SG.PE-4 | Moderate: SG.PE-4 | High: SG.PE-4 |

SG.PE-5 Visitor Control

Category: Common Governance, Risk, and Compliance (GRC) Requirements

Requirement

The organization controls physical access to the Smart Grid information system by authenticating visitors before authorizing access to the facility.

Supplemental Guidance

Contractors and others with permanent authorization credentials are not considered visitors.

Requirement Enhancements

The organization escorts visitors and monitors visitor activity as required according to security policies and procedures.

Additional Considerations

A1. The organization requires multiple forms of identification for access to the facility.

Impact Level Allocation

| Low: SG.PE-5 | Moderate: SG.PE-5 (1) | High: SG.PE-5 (1) |

SG.PE-6 Visitor Records

Category: Common Governance, Risk, and Compliance (GRC) Requirements

Requirement

The organization maintains visitor access records to the facility that include:

1. Name and organization of the person visiting;
2. Signature of the visitor;
3. Form of identification;
4. Date of access;
5. Time of entry and departure;
6. Purpose of visit; and
7. Name and organization of person visited.

Designated officials within the organization review the access logs after closeout and periodically review access logs based on an organization-defined frequency.

Supplemental Guidance

None.

Requirement Enhancements

None.

Additional Considerations

A1. The organization employs automated mechanisms to facilitate the maintenance and review of access records.

Impact Level Allocation

| Low: SG.PE-6 | Moderate: SG.PE-6 | High: SG.PE-6 |

SG.PE-7 Physical Access Log Retention

Category: Common Governance, Risk, and Compliance (GRC) Requirements

Requirement

The organization retains all physical access logs for as long as dictated by any applicable regulations or based on an organization-defined period by approved policy.

Supplemental Guidance

None.

Requirement Enhancements

None.

Additional Considerations

None.

Impact Level Allocation

| Low: SG.PE-7 | Moderate: SG.PE-7 | High: SG.PE-7 |

SG.PE-8   Emergency Shutoff Protection

Category: Common Governance, Risk, and Compliance (GRC) Requirements

Requirement

The organization protects the emergency power-off capability from accidental and intentional/unauthorized activation.

Supplemental Guidance

None.

Requirement Enhancements

None.

Additional Considerations

None.

Impact Level Allocation

| Low: SG.PE-8 | Moderate: SG.PE-8 | High: SG.PE-8 |

SG.PE-9   Emergency Power

Category: Common Governance, Risk, and Compliance (GRC) Requirements

Requirement

The organization provides an alternate power supply to facilitate an orderly shutdown of noncritical Smart Grid information system components in the event of a primary power source loss.

Supplemental Guidance

None.

Requirement Enhancements

1. The organization provides a long-term alternate power supply for the Smart Grid information system that is capable of maintaining minimally required operational capability in the event of an extended loss of the primary power source.

Additional Considerations

A1. The organization provides a long-term alternate power supply for the Smart Grid information system that is self-contained and not reliant on external power generation.

Impact Level Allocation

| Low: SG.PE-9 | Moderate: SG.PE-9 (1) | High: SG.PE-9 (1) |

SG.PE-10   Delivery and Removal

Category: Common Governance, Risk, and Compliance (GRC) Requirements

Requirement

The organization authorizes, monitors, and controls organization-defined types of Smart Grid information system components entering and exiting the facility and maintains records of those items.

Supplemental Guidance

The organization secures delivery areas and, if possible, isolates delivery areas from the Smart Grid information system to avoid unauthorized physical access.

Requirement Enhancements

None.

Additional Considerations

None.

Impact Level Allocation

| Low: SG.PE-10 | Moderate: SG.PE-10 | High: SG.PE-10 |

SG.PE-11   Alternate Work Site

Category: Common Governance, Risk, and Compliance (GRC) Requirements

Requirement

1. The organization establishes an alternate work site (for example, private residences) with proper equipment and communication infrastructure to compensate for the loss of the primary work site; and

2. The organization implements appropriate management, operational, and technical security measures at alternate control centers.

Supplemental Guidance

The organization may define different sets of security requirements for specific alternate work sites or types of sites.

Requirement Enhancements

None.

Additional Considerations

A1. The organization provides methods for employees to communicate with Smart Grid information system security staff in case of security problems.

Impact Level Allocation

| Low: SG.PE-11 | Moderate: SG.PE-11 | High: SG.PE-11 |

SG.PE-12 Location of Smart Grid Information System Assets

Category: Common Governance, Risk, and Compliance (GRC) Requirements

Requirement

The organization locates Smart Grid information system assets to minimize potential damage from physical and environmental hazards.

Supplemental Guidance

Physical and environmental hazards include flooding, fire, tornados, earthquakes, hurricanes, acts of terrorism, vandalism, electrical interference, and electromagnetic radiation.

Requirement Enhancements

1. The organization considers the risk associated with physical and environmental hazards when planning new Smart Grid information system facilities or reviewing existing facilities.

Additional Considerations

None.

Impact Level Allocation

| Low: SG.PE-12 | Moderate: SG.PE-12 | High: SG.PE-12 (1) |

3.19 PLANNING (SG.PL)

The purpose of strategic planning is to maintain optimal operations and to prevent or recover from undesirable interruptions to Smart Grid information system operation. Interruptions may take the form of a natural disaster (hurricane, tornado, earthquake, flood, etc.), an unintentional manmade event (accidental equipment damage, fire or explosion, operator error, etc.), an intentional manmade event (attack by bomb, firearm or vandalism, hacker or malware, etc.), or an equipment failure. The types of planning considered are security planning to prevent undesirable interruptions, continuity of operations planning to maintain Smart Grid information system operation during and after an interruption, and planning to identify mitigation strategies.

SG.PL-1 Strategic Planning Policy and Procedures

Category: Common Governance, Risk, and Compliance (GRC) Requirements

Requirement

1. The organization develops, implements, reviews, and updates on an organization-defined frequency—
   a. A documented planning policy that addresses—
      i. The objectives, roles, and responsibilities for the planning program as it relates to protecting the organization's personnel and assets; and
      ii. The scope of the planning program as it applies to all of the organizational staff, contractors, and third parties; and
   b. Procedures to address the implementation of the planning policy and associated strategic planning requirements;
2. Management commitment ensures compliance with the organization's security policy and other regulatory requirements; and
3. The organization ensures that the planning policy and procedures comply with applicable federal, state, local, tribal, and territorial laws and regulations.

Supplemental Guidance

The strategic planning policy may be included as part of the general information security policy for the organization. Strategic planning procedures may be developed for the security program in general and a Smart Grid information system in particular, when required.

Requirement Enhancements

None.

Additional Considerations

None.

Impact Level Allocation

Low: SG.PL-1	Moderate: SG.PL-1	High: SG.PL-1

SG.PL-2 Smart Grid Information System Security Plan

Category: Common Governance, Risk, and Compliance (GRC) Requirements

Requirement

The organization—

1. Develops a security plan for each Smart Grid information system that—
   a. Aligns with the organization's enterprise architecture;
   b. Explicitly defines the components of the Smart Grid information system;
   c. Describes relationships with and interconnections to other Smart Grid information systems;

d. Provides an overview of the security objectives for the Smart Grid information system;

e. Describes the security requirements in place or planned for meeting those requirements; and

f. Is reviewed and approved by the management authority prior to plan implementation;

2. Reviews the security plan for the Smart Grid information system on an organization-defined frequency; and

3. Revises the plan to address changes to the Smart Grid information system/environment of operation or problems identified during plan implementation or security requirement assessments.

Supplemental Guidance

None.

Requirement Enhancements

None.

Additional Considerations

None.

Impact Level Allocation

| Low: SG.PL-2 | Moderate: SG.PL-2 | High: SG.PL-2 |

SG.PL-3   Rules of Behavior

Category: Common Governance, Risk, and Compliance (GRC) Requirements

Requirement

The organization establishes and makes readily available to all Smart Grid information system users, a set of rules that describes their responsibilities and expected behavior with regard to Smart Grid information system usage.

Supplemental Guidance

None.

Requirement Enhancements

None.

Additional Considerations

A1. The organization includes in the rules of behavior, explicit restrictions on the use of social networking sites, posting information on commercial Web sites, and sharing Smart Grid information system account information; and

A2. The organization obtains signed acknowledgment from users indicating that they have read, understand, and agree to abide by the rules of behavior before authorizing access to the Smart Grid information system.

Impact Level Allocation

| Low: SG.PL-3 | Moderate: SG.PL-3 | High: SG.PL-3 |

SG.PL-4 Privacy Impact Assessment

Category: Common Governance, Risk, and Compliance (GRC) Requirements

Requirement

1. The organization conducts a privacy impact assessment on the Smart Grid information system; and
2. The privacy impact assessment is reviewed and approved by a management authority.

Supplemental Guidance

None.

Requirement Enhancements

None.

Additional Considerations

None.

Impact Level Allocation

| Low: SG.PL-4 | Moderate: SG.PL-4 | High: SG.PL-4 |

SG.PL-5 Security-Related Activity Planning

Category: Common Governance, Risk, and Compliance (GRC) Requirements

Requirement

1. The organization plans and coordinates security-related activities affecting the Smart Grid information system before conducting such activities to reduce the impact on organizational operations (i.e., mission, functions, image, and reputation), organizational assets, or individuals; and
2. Organizational planning and coordination includes both emergency and nonemergency (e.g., routine) situations.

Supplemental Guidance

Routine security-related activities include, but are not limited to, security assessments, audits, Smart Grid information system hardware, firmware, and software maintenance, and testing/exercises.

Requirement Enhancements

None.

Additional Considerations

None.

Impact Level Allocation

| Low: Not Selected | Moderate: SG.PL-5 | High: SG.PL-5 |

3.20 SECURITY PROGRAM MANAGEMENT (SG.PM)

The security program lays the groundwork for securing the organization's enterprise and Smart Grid information system assets. Security procedures define how an organization implements the security program.

SG.PM-1 Security Policy and Procedures

Category: Common Governance, Risk, and Compliance (GRC) Requirements

Requirement

1. The organization develops, implements, reviews, and updates on an organization-defined frequency—
    a. A documented security program security policy that addresses—
        i. The objectives, roles, and responsibilities for the security program as it relates to protecting the organization's personnel and assets; and
        ii. The scope of the security program as it applies to all of the organizational staff, contractors, and third parties; and
    b. Procedures to address the implementation of the security program security policy and associated security program protection requirements;
2. Management commitment ensures compliance with the organization's security policy and other regulatory requirements; and
3. The organization ensures that the security program security policy and procedures comply with applicable federal, state, local, tribal, and territorial laws and regulations.

Supplemental Guidance

The information system security policy can be included as part of the general security policy for the organization. Procedures can be developed for the security program in general and for the information system in particular, when required.

Requirement Enhancements

None.

Additional Considerations

None.

Impact Level Allocation

| Low: SG.PM-1 | Moderate: SG.PM-1 | High: SG.PM-1 |

SG.PM-2 Security Program Plan

Category: Common Governance, Risk, and Compliance (GRC) Requirements

Requirement

1. The organization develops and disseminates an organization-wide security program plan that—
   a. Provides an overview of the requirements for the security program and a description of the security program management requirements in place or planned for meeting those program requirements;
   b. Provides sufficient information about the program management requirements to enable an implementation that is compliant with the intent of the plan and a determination of the risk to be incurred if the plan is implemented as intended;
   c. Includes roles, responsibilities, management accountability, coordination among organizational entities, and compliance; and
   d. Is approved by a management authority with responsibility and accountability for the risk being incurred to organizational operations (including mission, functions, image, and reputation), organizational assets, and individuals;
2. Reviews the organization-wide security program plan on an organization-defined frequency; and
3. Revises the plan to address organizational changes and problems identified during plan implementation or security requirement assessments.

Supplemental Guidance

The security program plan documents the organization-wide program management requirements. The security plans for individual information systems and the organization-wide security program plan together, provide complete coverage for all security requirements employed within the organization.

Requirement Enhancements

None.

Additional Considerations

None.

Impact Level Allocation

| Low: SG.PM-2 | Moderate: SG.PM-2 | High: SG.PM-2 |

SG.PM-3   Senior Management Authority

Category: Common Governance, Risk, and Compliance (GRC) Requirements

Requirement

The organization appoints a senior management authority with the responsibility for the mission and resources to coordinate, develop, implement, and maintain an organization-wide security program.

Supplemental Guidance

None.

Requirement Enhancements

None.

Additional Considerations

None.

Impact Level Allocation

| Low: SG.PM-3 | Moderate: SG.PM-3 | High: SG.PM-3 |

SG.PM-4 Security Architecture

Category: Common Governance, Risk, and Compliance (GRC) Requirements

Requirement

The organization develops a security architecture with consideration for the resulting risk to organizational operations, organizational assets, individuals, and other organizations.

Supplemental Guidance

The integration of security requirements into the organization's enterprise architecture helps to ensure that security considerations are addressed by organizations early in the information system development life cycle.

Requirement Enhancements

None.

Additional Considerations

None.

Impact Level Allocation

| Low: SG.PM-4 | Moderate: SG.PM-4 | High: SG.PM-4 |

SG.PM-5 Risk Management Strategy

Category: Common Governance, Risk, and Compliance (GRC) Requirements

Requirement

The organization—

1. Develops a comprehensive strategy to manage risk to organizational operations and assets, individuals, and other organizations associated with the operation and use of information systems; and
2. Implements that strategy consistently across the organization.

Supplemental Guidance

An organization-wide risk management strategy should include a specification of the risk tolerance of the organization, guidance on acceptable risk assessment methodologies, and a process for consistently evaluating risk across the organization.

Requirement Enhancements

None.

Additional Considerations

None.

Impact Level Allocation

| Low: SG.PM-5 | Moderate: SG.PM-5 | High: SG.PM-5 |

SG.PM-6 Security Authorization to Operate Process

Category: Common Governance, Risk, and Compliance (GRC) Requirements

Requirement

The organization—

1. Manages (e.g., documents, tracks, and reports) the security state of organizational information systems through security authorization processes; and
2. Fully integrates the security authorization to operate processes into an organization-wide risk management strategy.

Supplemental Guidance

None.

Requirement Enhancements

None.

Impact Level Allocation

| Low: SG.PM-6 | Moderate: SG.PM-6 | High: SG.PM-6 |

SG.PM-7 Mission/Business Process Definition

Category: Common Governance, Risk, and Compliance (GRC) Requirements

Requirement

The organization defines mission/business processes that include consideration for security and the resulting risk to organizational operations, organizational assets, and individuals.

Supplemental Guidance

None.

Requirement Enhancements

None.

Additional Considerations

None.

Impact Level Allocation

| Low: SG.PM-7 | Moderate: SG.PM-7 | High: SG.PM-7 |

SG.PM-8 Management Accountability

Category: Common Governance, Risk, and Compliance (GRC) Requirements

Requirement

The organization defines a framework of management accountability that establishes roles and responsibilities to approve cyber security policy, assign security roles, and coordinate the implementation of cyber security across the organization.

Supplemental Guidance

None.

Requirement Enhancements

None.

Additional Considerations

None.

Impact Level Allocation

Low: SG.PM-8	Moderate: SG.PM-8	High: SG.PM-8

3.21 PERSONNEL SECURITY (SG.PS)

Personnel security addresses security program roles and responsibilities implemented during all phases of staff employment, including staff recruitment and termination. The organization screens applicants for critical positions in the operation and maintenance of the Smart Grid information system. The organization may consider implementing a confidentiality or nondisclosure agreement that employees and third-party users of facilities must sign before being granted access to the Smart Grid information system. The organization also documents and implements a process to secure resources and revoke access privileges when personnel terminate.

SG.PS-1 Personnel Security Policy and Procedures

Category: Common Governance, Risk, and Compliance (GRC) Requirements

Requirement

1. The organization develops, implements, reviews, and updates on an organization-defined frequency—
   a. A documented personnel security policy that addresses—
      i. The objectives, roles, and responsibilities for the personnel security program as it relates to protecting the organization's personnel and assets; and
      ii. The scope of the personnel security program as it applies to all of the organizational staff, contractors, and third parties; and
   b. Procedures to address the implementation of the personnel security policy and associated personnel protection requirements;
2. Management commitment ensures compliance with the organization's security policy and other regulatory requirements; and

3. The organization ensures that the personnel security policy and procedures comply with applicable federal, state, local, tribal, and territorial laws and regulations.

Supplemental Guidance

The personnel security policy may be included as part of the general information security policy for the organization. Personnel security procedures can be developed for the security program in general and for a particular Smart Grid information system, when required.

Requirement Enhancements

None.

Additional Considerations

None.

Impact Level Allocation

| Low: SG.PS-1 | Moderate: SG.PS-1 | High: SG.PS-1 |

SG.PS-2   Position Categorization

Category: Common Governance, Risk, and Compliance (GRC) Requirements

Requirement

The organization assigns a risk designation to all positions and establishes screening criteria for individuals filling those positions. The organization reviews and revises position risk designations. The organization determines the frequency of the review based on the organization's requirements or regulatory commitments.

Supplemental Guidance

None.

Requirement Enhancements

None.

Additional Considerations

None.

Impact Level Allocation

| Low: SG.PS-2 | Moderate: SG.PS-2 | High: SG.PS-2 |

SG.PS-3   Personnel Screening

Category: Common Governance, Risk, and Compliance (GRC) Requirements

Requirement

The organization screens individuals requiring access to the Smart Grid information system before access is authorized. The organization maintains consistency between the screening process and organization-defined policy, regulations, guidance, and the criteria established for the risk designation of the assigned position.

Supplemental Guidance

Basic screening requirements should include:

1. Employment history;
2. Verification of the highest education degree received;
3. Residency;
4. References; and
5. Law enforcement records.

Requirement Enhancements

None.

Additional Considerations

A1. The organization rescreens individuals with access to Smart Grid information systems based on a defined list of conditions requiring rescreening and the frequency of such rescreening.

Impact Level Allocation

| Low: SG.PS-3 | Moderate: SG.PS-3 | High: SG.PS-3 |

SG.PS-4   Personnel Termination

Category: Common Governance, Risk, and Compliance (GRC) Requirements

Requirement

1. When an employee is terminated, the organization revokes logical and physical access to facilities and systems and ensures that all organization-owned property is returned. Organization-owned documents relating to the Smart Grid information system that are in the employee's possession are transferred to the new authorized owner;
2. All logical and physical access must be terminated at an organization-defined time frame for personnel terminated for cause; and
3. Exit interviews ensure that individuals understand any security constraints imposed by being a former employee and that proper accountability is achieved for all Smart Grid information system-related property.

Supplemental Guidance

Organization-owned property includes Smart Grid information system administration manuals, keys, identification cards, building passes, computers, cell phones, and personal data assistants. Organization-owned documents include field device configuration and operational information and Smart Grid information system network documentation.

Requirement Enhancements

None.

Additional Considerations

A1. The organization implements automated processes to revoke access permissions that are initiated by the termination.

Impact Level Allocation

Low: SG.PS-4	Moderate: SG.PS-4	High: SG.PS-4

SG.PS-5 Personnel Transfer

Category: Common Governance, Risk, and Compliance (GRC) Requirements

Requirement

1. The organization reviews logical and physical access permissions to Smart Grid information systems and facilities when individuals are reassigned or transferred to other positions within the organization and initiates appropriate actions; and
2. Complete execution of this requirement occurs within an organization-defined time period for employees, contractors, or third parties who no longer need to access Smart Grid information system resources.

Supplemental Guidance

Appropriate actions may include:

1. Returning old and issuing new keys, identification cards, and building passes;
2. Closing old accounts and establishing new accounts;
3. Changing Smart Grid information system access authorizations; and
4. Providing access to official records created or managed by the employee at the former work location and in the former accounts.

Requirement Enhancements

None.

Additional Considerations

None.

Impact Level Allocation

Low: SG.PS-5	Moderate: SG.PS-5	High: SG.PS-5

SG.PS-6 Access Agreements

Category: Common Governance, Risk, and Compliance (GRC) Requirements

Requirement

1. The organization completes appropriate agreements for Smart Grid information system access before access is granted. This requirement applies to all parties, including third parties and contractors, who require access to the Smart Grid information system;
2. The organization reviews and updates access agreements periodically; and

3. Signed access agreements include an acknowledgment that individuals have read, understand, and agree to abide by the constraints associated with the Smart Grid information system to which access is authorized.

Supplemental Guidance

Access agreements include nondisclosure agreements, acceptable use agreements, rules of behavior, and conflict-of-interest agreements.

Requirement Enhancements

None.

Additional Considerations

None.

Impact Level Allocation

| Low: SG.PS-6 | Moderate: SG.PS-6 | High: SG.PS-6 |

SG.PS-7   Contractor and Third-Party Personnel Security

Category: Common Governance, Risk, and Compliance (GRC) Requirements

Requirement

The organization enforces security requirements for contractor and third-party personnel and monitors service provider behavior and compliance.

Supplemental Guidance

Contactors and third-party providers include service bureaus and other organizations providing Smart Grid information system operation and maintenance, development, IT services, outsourced applications, and network and security management.

Requirement Enhancements

None.

Additional Considerations

None.

Impact Level Allocation

| Low: SG.PS-7 | Moderate: SG.PS-7 | High: SG.PS-7 |

SG.PS-8   Personnel Accountability

Category: Common Governance, Risk, and Compliance (GRC) Requirements

Requirement

1. The organization employs a formal accountability process for personnel failing to comply with established security policies and procedures and identifies disciplinary actions for failing to comply; and

2. The organization ensures that the accountability process complies with applicable federal, state, local, tribal, and territorial laws and regulations.

Supplemental Guidance

The accountability process can be included as part of the organization's general personnel policies and procedures.

Requirement Enhancements

None.

Additional Considerations

None.

Impact Level Allocation

| Low: SG.PS-8 | Moderate: SG.PS-8 | High: SG.PS-8 |

SG.PS-9 Personnel Roles

Category: Common Governance, Risk, and Compliance (GRC) Requirements

Requirement

The organization provides employees, contractors, and third parties with expectations of conduct, duties, terms and conditions of employment, legal rights, and responsibilities.

Supplemental Guidance

None.

Requirement Enhancements

None.

Additional Considerations

A1. Employees and contractors acknowledge understanding by signature.

Impact Level Allocation

| Low: SG.PS-9 | Moderate: SG.PS-9 | High: SG.PS-9 |

3.22 RISK MANAGEMENT AND ASSESSMENT (SG.RA)

Risk management planning is a key aspect of ensuring that the processes and technical means of securing Smart Grid information systems have fully addressed the risks and vulnerabilities in the Smart Grid information system.

An organization identifies and classifies risks to develop appropriate security measures. Risk identification and classification involves security assessments of Smart Grid information systems and interconnections to identify critical components and any areas weak in security. The risk identification and classification process is continually performed to monitor the Smart Grid information system's compliance status.

SG.RA-1 Risk Assessment Policy and Procedures

Category: Common Governance, Risk, and Compliance (GRC) Requirements

Requirement

1. The organization develops, implements, reviews, and updates on an organization-defined frequency—
   a. A documented risk assessment security policy that addresses—
      i. The objectives, roles, and responsibilities for the risk assessment security program as it relates to protecting the organization's personnel and assets; and
      ii. The scope of the risk assessment security program as it applies to all of the organizational staff, contractors, and third parties; and
   b. Procedures to address the implementation of the risk assessment security policy and associated risk assessment protection requirements;
2. Management commitment ensures compliance with the organization's security policy and other regulatory requirements; and
3. The organization ensures that the risk assessment policy and procedures comply with applicable federal, state, local, tribal, and territorial laws and regulations.

Supplemental Guidance

The risk assessment policy also takes into account the organization's risk tolerance level. The risk assessment policy can be included as part of the general security policy for the organization. Risk assessment procedures can be developed for the security program in general and for a particular Smart Grid information system, when required.

Requirement Enhancements

None.

Additional Considerations

None.

Impact Level Allocation

Low: SG.RA-1	Moderate: SG.RA-1	High: SG.RA-1

SG.RA-2 Risk Management Plan

Category: Common Governance, Risk, and Compliance (GRC) Requirements

Requirement

1. The organization develops a risk management plan;
2. A management authority reviews and approves the risk management plan; and
3. Risk-reduction mitigation measures are planned and implemented and the results monitored to ensure effectiveness of the organization's risk management plan.

Supplemental Guidance

Risk mitigation measures need to be implemented and the results monitored against planned metrics to ensure the effectiveness of the risk management plan.

Requirement Enhancements

None.

Additional Considerations

None.

Impact Level Allocation

Low: SG.RA-2	Moderate: SG.RA-2	High: SG.RA-2

SG.RA-3   Security Impact Level

Category: Common Governance, Risk, and Compliance (GRC) Requirements

Requirement

The organization—

1. Specifies the information and the information system impact levels;
2. Documents the impact level results (including supporting rationale) in the security plan for the information system; and
3. Reviews the Smart Grid information system and information impact levels on an organization-defined frequency.

Supplemental Guidance

Impact level designation is based on the need, priority, and level of protection required commensurate with sensitivity and impact of the loss of availability, integrity, or confidentiality. Impact level designation may also be based on regulatory requirements, for example, the NERC CIPs. The organization considers safety issues in determining the impact level for the Smart Grid information system.

Requirement Enhancements

None.

Additional Considerations

None.

Impact Level Allocation

Low: SG.RA-3	Moderate: SG.RA-3	High: SG.RA-3

SG.RA-4   Risk Assessment

Category: Common Governance, Risk, and Compliance (GRC) Requirements

Requirement

The organization—

1. Conducts assessments of risk from the unauthorized access, use, disclosure, disruption, modification, or destruction of information and Smart Grid information systems; and

2. Updates risk assessments on an organization-defined frequency or whenever significant changes occur to the Smart Grid information system or environment of operation, or other conditions that may impact the security of the Smart Grid information system.

Supplemental Guidance

Risk assessments take into account vulnerabilities, threat sources, risk tolerance levels, and security mechanisms planned or in place to determine the resulting level of residual risk posed to organizational operations, organizational assets, or individuals based on the operation of the Smart Grid information system.

Requirement Enhancements

None.

Additional Considerations

None.

Impact Level Allocation

Low: SG.RA-4	Moderate: SG.RA-4	High: SG.RA-4

SG.RA-5 Risk Assessment Update

Category: Common Governance, Risk, and Compliance (GRC) Requirements

Requirement

The organization updates the risk assessment plan on an organization-defined frequency or whenever significant changes occur to the Smart Grid information system, the facilities where the Smart Grid information system resides, or other conditions that may affect the security or authorization-to-operate status of the Smart Grid information system.

Supplemental Guidance

The organization develops and documents specific criteria for what are considered significant changes to the Smart Grid information system.

Requirement Enhancements

None.

Additional Considerations

None.

Impact Level Allocation

Low: SG.RA-5	Moderate: SG.RA-5	High: SG.RA-5

SG.RA-6 Vulnerability Assessment and Awareness

Category: Common Governance, Risk, and Compliance (GRC) Requirements

Requirement

The organization—

1. Monitors and evaluates the Smart Grid information system according to the risk management plan on an organization-defined frequency to identify vulnerabilities that might affect the security of a Smart Grid information system;
2. Analyzes vulnerability scan reports and remediates vulnerabilities within an organization-defined time frame based on an assessment of risk;
3. Shares information obtained from the vulnerability scanning process with designated personnel throughout the organization to help eliminate similar vulnerabilities in other Smart Grid information systems;
4. Updates the Smart Grid information system to address any identified vulnerabilities in accordance with organization's Smart Grid information system maintenance policy; and
5. Updates the list of Smart Grid information system vulnerabilities on an organization-defined frequency or when new vulnerabilities are identified and reported.

Supplemental Guidance

Vulnerability analysis for custom software and applications may require additional, more specialized approaches (e.g., vulnerability scanning tools to scan for Web-based vulnerabilities, source code reviews, and static analysis of source code). Vulnerability scanning includes scanning for ports, protocols, and services that should not be accessible to users and for improperly configured or incorrectly operating information flow mechanisms.

Requirement Enhancements

1. The organization employs vulnerability scanning tools that include the capability to update the list of Smart Grid information system vulnerabilities scanned; and
2. The organization includes privileged access authorization to organization-defined Smart Grid information system components for selected vulnerability scanning activities to facilitate more thorough scanning.

Additional Considerations

A1. The organization employs automated mechanisms on an organization-defined frequency to detect the presence of unauthorized software on organizational Smart Grid information systems and notifies designated organizational officials;

A2. The organization performs security testing to determine the level of difficulty in circumventing the security requirements of the Smart Grid information system; and

A3. The organization employs automated mechanisms to compare the results of vulnerability scans over time to determine trends in Smart Grid information system vulnerabilities.

Impact Level Allocation

Low: SG.RA-6	Moderate: SG.RA-6 (1)	High: SG.RA-6 (1), (2)

3.23 SMART GRID INFORMATION SYSTEM AND SERVICES ACQUISITION (SG.SA)

Smart Grid information systems and services acquisition covers the contracting and acquiring of system components, software, firmware, and services from employees, contactors, and third parties. A policy with detailed procedures for reviewing acquisitions should reduce the introduction of additional or unknown vulnerabilities into the Smart Grid information system.

SG.SA-1 Smart Grid Information System and Services Acquisition Policy and Procedures

Category: Common Governance, Risk, and Compliance (GRC) Requirements

Requirement

1. The organization develops, implements, reviews, and updates on an organization-defined frequency—

    a. A documented Smart Grid information system and services acquisition security policy that addresses—

        i. The objectives, roles, and responsibilities for the Smart Grid information system and services acquisition security program as it relates to protecting the organization's personnel and assets; and

        ii. The scope of the Smart Grid information system and services acquisition security program as it applies to all of the organizational staff, contractors, and third parties; and

    b. Procedures to address the implementation of the Smart Grid information system and services acquisition policy and associated physical and environmental protection requirements;

2. Management commitment ensures compliance with the organization's security policy and other regulatory requirements; and

3. The organization ensures that the Smart Grid information system and services acquisition policy and procedures comply with applicable federal, state, local, tribal, and territorial laws and regulations.

Supplemental Guidance

The Smart Grid information system and services acquisition policy can be included as part of the general information security policy for the organization. Smart Grid information system and services acquisition procedures can be developed for the security program in general and for a particular Smart Grid information system when required.

Requirement Enhancements

None.

Additional Considerations

None.

Impact Level Allocation

| Low: SG.SA-1 | Moderate: SG.SA-1 | High: SG.SA-1 |

SG.SA-2   Security Policies for Contractors and Third Parties

Category: Common Governance, Risk, and Compliance (GRC) Requirements

Requirement

1. External suppliers and contractors that have an impact on the security of Smart Grid information systems must meet the organization's policy and procedures; and
2. The organization establishes procedures to remove external supplier and contractor access to Smart Grid information systems at the conclusion/termination of the contract.

Supplemental Guidance

The organization considers the increased security risk associated with outsourcing as part of the decision-making process to determine what to outsource and what outsourcing partner to select. Contracts with external suppliers govern physical as well as logical access. The organization considers confidentiality or nondisclosure agreements and intellectual property rights.

Requirement Enhancements

None.

Additional Considerations

None.

Impact Level Allocation

Low: SG.SA-2	Moderate: SG.SA-2	High: SG.SA-2

SG.SA-3   Life-Cycle Support

Category: Common Governance, Risk, and Compliance (GRC) Requirements

Requirement

The organization manages the Smart Grid information system using a system development lifecycle methodology that includes security.

Supplemental Guidance

None.

Requirement Enhancements

None.

Additional Considerations

None.

Impact Level Allocation

Low: SG.SA-3	Moderate: SG.SA-3	High: SG.SA-3

SG.SA-4   Acquisitions

Category: Common Governance, Risk, and Compliance (GRC) Requirements

Requirement

The organization includes security requirements in Smart Grid information system acquisition contracts in accordance with applicable laws, regulations, and organization-defined security policies.

Supplemental Guidance

None.

Requirement Enhancements

None.

Additional Considerations

None.

Impact Level Allocation

| Low: SG.SA-4 | Moderate: SG.SA-4 | High: SG.SA-4 |

SG.SA-5 Smart Grid Information System Documentation

Category: Common Governance, Risk, and Compliance (GRC) Requirement

Requirement

1. Smart Grid information system documentation includes how to configure, install, and use the information system and the information system's security features; and
2. The organization obtains from the contractor/third-party, information describing the functional properties of the security controls employed within the Smart Grid information system.

Supplemental Guidance

None.

Requirement Enhancements

None.

Additional Considerations

None.

Impact Level Allocation

| Low: SG.SA-5 | Moderate: SG.SA-5 | High: SG.SA-5 |

SG.SA-6 Software License Usage Restrictions

Category: Common Governance, Risk, and Compliance (GRC) Requirements

Requirement

The organization—

1. Uses software and associated documentation in accordance with contract agreements and copyright laws; and

2. Controls the use of software and associated documentation protected by quantity licenses and copyrighted material.

Supplemental Guidance

None.

Requirement Enhancements

None.

Additional Considerations

None.

Impact Level Allocation

| Low: SG.SA-6 | Moderate: SG.SA-6 | High: SG.SA-6 |

SG.SA-7 User-Installed Software

Category: Common Governance, Risk, and Compliance (GRC) Requirements

Requirement

The organization establishes policies and procedures to manage user installation of software.

Supplemental Guidance

If provided the necessary privileges, users have the ability to install software. The organization's security program identifies the types of software permitted to be downloaded and installed (e.g., updates and security patches to existing software) and types of software prohibited (e.g., software that is free only for personal, not corporate use, and software whose pedigree with regard to being potentially malicious is unknown or suspect).

Requirement Enhancements

None.

Additional Considerations

None.

Impact Level Allocation

| Low: SG.SA-7 | Moderate: SG.SA-7 | High: SG.SA-7 |

SG.SA-8 Security Engineering Principles

Category: Common Governance, Risk, and Compliance (GRC) Requirements

Requirement

The organization applies security engineering principles in the specification, design, development, and implementation of any Smart Grid information system.

Security engineering principles include:

1. Ongoing secure development education requirements for all developers involved in the Smart Grid information system;

2. Specification of a minimum standard for security;
3. Specification of a minimum standard for privacy;
4. Creation of a threat model for a Smart Grid information system;
5. Updating of product specifications to include mitigations for threats discovered during threat modeling;
6. Use of secure coding practices to reduce common security errors;
7. Testing to validate the effectiveness of secure coding practices;
8. Performance of a final security audit prior to authorization to operate to confirm adherence to security requirements;
9. Creation of a documented and tested security response plan in the event vulnerability is discovered;
10. Creation of a documented and tested privacy response plan in the event vulnerability is discovered; and
11. Performance of a root cause analysis to understand the cause of identified vulnerabilities.

Supplemental Guidance

The application of security engineering principles is primarily targeted at new development Smart Grid information systems or Smart Grid information systems undergoing major upgrades. These principles are integrated into the Smart Grid information system development life cycle. For legacy Smart Grid information systems, the organization applies security engineering principles to Smart Grid information system upgrades and modifications, to the extent feasible, given the current state of the hardware, software, and firmware components within the Smart Grid information system. The organization minimizes risk to legacy systems through attack surface reduction and other mitigating controls.

Requirement Enhancements

None.

Additional Considerations

None.

Impact Level Allocation

Low: SG.SA-8	Moderate: SG.SA-8	High: SG.SA-8

SG.SA-9   Developer Configuration Management

Category: Common Governance, Risk, and Compliance (GRC) Requirements

Requirement

The organization requires that Smart Grid information system developers/integrators document and implement a configuration management process that—

1. Manages and controls changes to the Smart Grid information system during design, development, implementation, and operation;

2. Tracks security flaws; and
3. Includes organizational approval of changes.

Supplemental Guidance

None.

Requirement Enhancements

None.

Additional Considerations

A1. The organization requires that Smart Grid information system developers/integrators provide an integrity check of delivered software and firmware.

Impact Level Allocation

| Low: SG.SA-9 | Moderate: SG.SA-9 | High: SG.SA-9 |

SG.SA-10 Developer Security Testing

Category: Common Technical Requirements, Integrity

Requirement

1. The Smart Grid information system developer creates a security test and evaluation plan;
2. The developer submits the plan to the organization for approval and implements the plan once written approval is obtained;
3. The developer documents the results of the testing and evaluation and submits them to the organization for approval; and
4. The organization does not perform developmental security tests on the production Smart Grid information system.

Supplemental Guidance

None.

Requirement Enhancements

None.

Additional Considerations

A1. The organization requires that Smart Grid information system developers employ code analysis tools to examine software for common flaws and document the results of the analysis; and

A2. The organization requires that Smart Grid information system developers/integrators perform a vulnerability analysis to document vulnerabilities, exploitation potential, and risk mitigations.

Impact Level Allocation

| Low: SG.SA-10 | Moderate: SG.SA-10 | High: SG.SA-10 |

SG.SA-11   Supply Chain Protection

Category: Common Technical Requirements, Integrity

Requirement

The organization protects against supply chain vulnerabilities employing requirements defined to protect the products and services from threats initiated against organizations, people, information, and resources, possibly international in scope, that provides products or services to the organization.

Supplemental Guidance

Supply chain protection helps to protect Smart Grid information systems (including the technology products that compose those Smart Grid information systems) throughout the system development life cycle (e.g., during design and development, manufacturing, packaging, assembly, distribution, system integration, operations, maintenance, and retirement).

Requirement Enhancements

None.

Additional Considerations

A1.   The organization conducts a due diligence review of suppliers prior to entering into contractual agreements to acquire Smart Grid information system hardware, software, firmware, or services;

A2.   The organization uses a diverse set of suppliers for Smart Grid information systems, Smart Grid information system components, technology products, and Smart Grid information system services; and

A3.   The organization employs independent analysis and penetration testing against delivered Smart Grid information systems, Smart Grid information system components, and technology products.

Impact Level Allocation

Low: SG.SA-11	Moderate: SG.SA-11	High: SG.SA-11

3.24 SMART GRID INFORMATION SYSTEM AND COMMUNICATION PROTECTION (SG.SC)

Smart Grid information system and communication protection consists of steps taken to protect the Smart Grid information system and the communication links between Smart Grid information system components from cyber intrusions. Although Smart Grid information system and communication protection might include both physical and cyber protection, this section addresses only cyber protection. Physical protection is addressed in SG.PE, Physical and Environmental Security.

SG.SC-1   Smart Grid Information System and Communication Protection Policy and Procedures

Category: Common Governance, Risk, and Compliance (GRC) Requirements

Requirement

1. The organization develops, implements, reviews, and updates on an organization-defined frequency—

    a. A documented Smart Grid information system and communication protection security policy that addresses—

        i. The objectives, roles, and responsibilities for the Smart Grid information system and communication protection security program as it relates to protecting the organization's personnel and assets; and

        ii. The scope of the Smart Grid information system and communication protection policy as it applies to all of the organizational staff, contractors, and third parties; and

    b. Procedures to address the implementation of the Smart Grid information system and communication protection security policy and associated Smart Grid information system and communication protection requirements;

2. Management commitment ensures compliance with the organization's security policy and other regulatory requirements; and

3. The organization ensures that the Smart Grid information system and communication protection policy and procedures comply with applicable federal, state, local, tribal, and territorial laws and regulations.

Supplemental Guidance

The Smart Grid information system and communication protection policy may be included as part of the general information security policy for the organization. Smart Grid information system and communication protection procedures can be developed for the security program in general and a Smart Grid information system in particular, when required.

Requirement Enhancements

None.

Additional Considerations

None.

Impact Level Allocation

Low: SG.SC-1	Moderate: SG.SC-1	High: SG.SC-1

SG.SC-2   Communications Partitioning

Category: Unique Technical Requirements

Requirement

The Smart Grid information system partitions the communications for telemetry/data acquisition services and management functionality.

Supplemental Guidance

The Smart Grid information system management communications path needs to be physically or logically separated from the telemetry/data acquisition services communications path.

Requirement Enhancements

None.

Additional Considerations

None.

Impact Level Allocation

| Low: Not Selected | Moderate: Not Selected | High: Not Selected |

SG.SC-3   Security Function Isolation

Category: Unique Technical Requirements

Requirement

The Smart Grid information system isolates security functions from nonsecurity functions.

Supplemental Guidance

None.

Requirement Enhancements

None.

Additional Considerations

A1.   The Smart Grid information system employs underlying hardware separation mechanisms to facilitate security function isolation; and

A2.   The Smart Grid information system isolates security functions (e.g., functions enforcing access and information flow control) from both nonsecurity functions and from other security functions.

Impact Level Allocation

| Low: SG.SC-3 | Moderate: SG.SC-3 | High: SG.SC-3 |

SG.SC-4   Information Remnants

Category: Unique Technical Requirements

Requirement

The Smart Grid information system prevents unauthorized or unintended information transfer via shared Smart Grid information system resources.

Supplemental Guidance

Control of Smart Grid information system remnants, sometimes referred to as object reuse, or data remnants, prevents information from being available to any current user/role/process that obtains access to a shared Smart Grid information system resource after that resource has been released back to the Smart Grid information system.

Requirement Enhancements

None.

Additional Considerations

None.

Impact Level Allocation

| Low: Not Selected | Moderate: SG.SC-4 | High: SG.SC-4 |

SG.SC-5   Denial-of-Service Protection

Category: Unique Technical Requirements

Requirement

The Smart Grid information system mitigates or limits the effects of denial-of-service attacks based on an organization-defined list of denial-of-service attacks.

Supplemental Guidance

Network perimeter devices can filter certain types of packets to protect devices on an organization's internal network from being directly affected by denial-of-service attacks.

Requirement Enhancements

None.

Additional Considerations

A1.   The Smart Grid information system restricts the ability of users to launch denial-of-service attacks against other Smart Grid information systems or networks; and

A2.   The Smart Grid information system manages excess capacity, bandwidth, or other redundancy to limit the effects of information flooding types of denial-of-service attacks.

Impact Level Allocation

| Low: SG.SC-5 | Moderate: SG.SC-5 | High: SG.SC-5 |

SG.SC-6   Resource Priority

Category: Unique Technical Requirements

Requirement

The Smart Grid information system prioritizes the use of resources.

Supplemental Guidance

Priority protection helps prevent a lower-priority process from delaying or interfering with the Smart Grid information system servicing any higher-priority process. This requirement does not apply to components in the Smart Grid information system for which only a single user/role exists.

Requirement Enhancements

None.

Additional Considerations

None.

Impact Level Allocation

| Low: Not Selected | Moderate: Not Selected | High: Not Selected |

SG.SC-7 Boundary Protection

Category: Unique Technical Requirements

Requirement

1. The organization defines the boundary of the Smart Grid information system;
2. The Smart Grid information system monitors and controls communications at the external boundary of the system and at key internal boundaries within the system;
3. The Smart Grid information system connects to external networks or information systems only through managed interfaces consisting of boundary protection devices;
4. The managed interface implements security measures appropriate for the protection of integrity and confidentiality of the transmitted information; and
5. The organization prevents public access into the organization's internal Smart Grid information system networks except as appropriately mediated.

Supplemental Guidance

Managed interfaces employing boundary protection devices include proxies, gateways, routers, firewalls, guards, or encrypted tunnels.

Requirement Enhancements

1. The Smart Grid information system denies network traffic by default and allows network traffic by exception (i.e., deny all, permit by exception);
2. The Smart Grid information system checks incoming communications to ensure that the communications are coming from an authorized source and routed to an authorized destination; and
3. Communications to/from Smart Grid information system components shall be restricted to specific components in the Smart Grid information system. Communications shall not be permitted to/from any non-Smart Grid system unless separated by a controlled logical/physical interface.

Additional Considerations

A1. The organization prevents the unauthorized release of information outside the Smart Grid information system boundary or any unauthorized communication through the Smart Grid information system boundary when an operational failure occurs of the boundary protection mechanisms;

A2. The organization prevents the unauthorized exfiltration of information across managed interfaces;

A3. The Smart Grid information system routes internal communications traffic to the Internet through authenticated proxy servers within the managed interfaces of boundary protection devices;

A4. The organization limits the number of access points to the Smart Grid information system to allow for better monitoring of inbound and outbound network traffic;

A5. Smart Grid information system boundary protections at any designated alternate processing/control sites provide the same levels of protection as that of the primary site; and

A6. The Smart Grid information system fails securely in the event of an operational failure of a boundary protection device.

Impact Level Allocation

Low: SG.SC-7	Moderate: SG.SC-7 (1), (2), (3)	High: SG.SC-7 (1), (2), (3)

SG.SC-8 Communication Integrity

Category: Unique Technical Requirements

Requirement

The Smart Grid information system protects the integrity of electronically communicated information.

Supplemental Guidance

None.

Requirement Enhancements

1. The organization employs cryptographic mechanisms to ensure integrity.

Additional Considerations

A1. The Smart Grid information system maintains the integrity of information during aggregation, packaging, and transformation in preparation for transmission.

Impact Level Allocation

Low: Not Selected	Moderate: SG.SC-8 (1)	High: SG.SC-8 (1)

SG.SC-9 Communication Confidentiality

Category: Unique Technical Requirements

Requirement

The Smart Grid information system protects the confidentiality of communicated information.

Supplemental Guidance

None.

Requirement Enhancements

1. The organization employs cryptographic mechanisms to prevent unauthorized disclosure of information during transmission.

Additional Considerations

None.

Impact Level Allocation

| Low: Not Selected | Moderate: SG.SC-9 (1) | High: SG.SC-9 (1) |

SG.SC-10   Trusted Path

Category: Unique Technical Requirements

Requirement

The Smart Grid information system establishes a trusted communications path between the user and the Smart Grid information system.

Supplemental Guidance

A trusted path is the means by which a user and target of evaluation security functionality can communicate with the necessary confidence.

Requirement Enhancements

None.

Additional Considerations

None.

Impact Level Allocation

| Low: Not Selected | Moderate: Not Selected | High: Not Selected |

SG.SC-11   Cryptographic Key Establishment and Management

Category: Common Technical Requirements, Confidentiality

Requirement

The organization establishes and manages cryptographic keys for required cryptography employed within the information system.

Supplemental Guidance

Key establishment includes a key generation process in accordance with a specified algorithm and key sizes, and key sizes based on an assigned standard. Key generation must be performed using an appropriate random number generator. The policies for key management need to address such items as periodic key changes, key destruction, and key distribution.

Requirement Enhancements

1. The organization maintains availability of information in the event of the loss of cryptographic keys by users. *See* Chapter 4 for key management requirements.

Additional Considerations

None.

Impact Level Allocation

| Low: SG.SC-11 | Moderate: SG.SC-11 (1) | High: SG.SC-11 (1) |

SG.SC-12   Use of Validated Cryptography

Category: Common Technical Requirements, Confidentiality

Requirement

All of the cryptography and other security functions (e.g., hashes, random number generators, etc.) that are required for use in a Smart Grid information system shall be NIST Federal Information Processing Standard (FIPS) approved or allowed for use in FIPS modes.

Supplemental Guidance

For a list of current FIPS-approved or allowed cryptography, *see* Chapter Four Cryptography and Key Management.

Requirement Enhancements

None.

Additional Considerations

None.

Impact Level Allocation

| Low: SG.SC-12 | Moderate: SG.SC-12 | High: SG.SC-12 |

SG.SC-13   Collaborative Computing

Category: Common Governance, Risk, and Compliance (GRC) Requirements

Requirement

The organization develops, disseminates, and periodically reviews and updates on an organization-defined frequency a collaborative computing policy.

Supplemental Guidance

Collaborative computing mechanisms include video and audio conferencing capabilities or instant messaging technologies. Explicit indication of use includes signals to local users when cameras and/or microphones are activated.

Requirement Enhancements

None.

Additional Considerations

None.

Impact Level Allocation

| Low: SG.SC-13 | Moderate: SG.SC-13 | High: SG.SC-13 |

SG.SC-14 Transmission of Security Parameters

Category: Unique Technical Requirements

Requirement

The Smart Grid information system reliably associates security parameters with information exchanged between the enterprise information systems and the Smart Grid information system.

Supplemental Guidance

Security parameters may be explicitly or implicitly associated with the information contained within the Smart Grid information system.

Requirement Enhancements

None.

Additional Considerations

A1. The Smart Grid information system validates the integrity of security parameters exchanged between Smart Grid information systems.

Impact Level Allocation

Low: Not Selected	Moderate: Not Selected	High: Not Selected

SG.SC-15 Public Key Infrastructure Certificates

Category: Common Technical Requirements, Confidentiality

Requirement

For Smart Grid information systems that implement a public key infrastructure, the organization issues public key certificates under an appropriate certificate policy or obtains public key certificates under an appropriate certificate policy from an approved service provider.

Supplemental Guidance

Registration to receive a public key certificate needs to include authorization by a supervisor or a responsible official and needs to be accomplished using a secure process that verifies the identity of the certificate holder and ensures that the certificate is issued to the intended party.

Requirement Enhancements

None.

Additional Considerations

None.

Impact Level Allocation

Low: SG.SC-15	Moderate: SG.SC-15	High: SG.SC-15

SG.SC-16 Mobile Code

Category: Common Technical Requirements, Confidentiality

Requirement

The organization—

1. Establishes usage restrictions and implementation guidance for mobile code technologies based on the potential to cause damage to the Smart Grid information system if used maliciously;
2. Documents, monitors, and manages the use of mobile code within the Smart Grid information system; and
3. A management authority authorizes the use of mobile code.

Supplemental Guidance

Mobile code technologies include, for example, Java, JavaScript, ActiveX, PDF, Postscript, Shockwave movies, Flash animations, and VBScript. Usage restrictions and implementation guidance need to apply to both the selection and use of mobile code installed on organizational servers and mobile code downloaded and executed on individual workstations.

Requirement Enhancements

None.

Additional Considerations

A1. The Smart Grid information system implements detection and inspection mechanisms to identify unauthorized mobile code and takes corrective actions, when necessary.

Impact Level Allocation

Low: Not Selected	Moderate: SG.SC-16	High: SG.SC-16

SG.SC-17   Voice-Over Internet Protocol

Category: Unique Technical Requirements

Requirement

The organization—

1. Establishes usage restrictions and implementation guidance for VoIP technologies based on the potential to cause damage to the Smart Grid information system if used maliciously; and
2. Authorizes, monitors, and controls the use of VoIP within the Smart Grid information system.

Supplemental Guidance

None.

Requirement Enhancements

None.

Additional Considerations

None.

Impact Level Allocation

| Low: Not Selected | Moderate: SG.SC-17 | High: SG.SC-17 |

SG.SC-18   System Connections

Category: Common Technical Requirements, Confidentiality

Requirement

All external Smart Grid information system and communication connections are identified and protected from tampering or damage.

Supplemental Guidance

External access point connections to the Smart Grid information system need to be secured to protect the Smart Grid information system. Access points include any externally connected communication end point (for example, dial-up modems).

Requirement Enhancements

None.

Additional Considerations

None.

Impact Level Allocation

| Low: SG.SC-18 | Moderate: SG.SC-18 | High: SG.SC-18 |

SG.SC-19   Security Roles

Category: Common Technical Requirements, Integrity

Requirement

The Smart Grid information system design and implementation specifies the security roles and responsibilities for the users of the Smart Grid information system.

Supplemental Guidance

Security roles and responsibilities for Smart Grid information system users need to be specified, defined, and implemented based on the sensitivity of the information handled by the user. These roles may be defined for specific job descriptions or for individuals.

Requirement Enhancements

None.

Additional Considerations

None.

Impact Level Allocation

| Low: SG.SC-19 | Moderate: SG.SC-19 | High: SG.SC-19 |

SG.SC-20   Message Authenticity

Category: Common Technical Requirements, Integrity

Requirement

The Smart Grid information system provides mechanisms to protect the authenticity of device-to-device communications.

Supplemental Guidance

Message authentication provides protection from malformed traffic, misconfigured devices, and malicious entities.

Requirement Enhancements

None.

Additional Considerations

A1.   Message authentication mechanisms should be implemented at the protocol level for both serial and routable protocols.

Impact Level Allocation

Low: SG.SC-20	Moderate: SG.SC-20	High: SG.SC-20

SG.SC-21   Secure Name/Address Resolution Service

Category: Common Technical Requirements, Integrity

Requirement

The organization is responsible for—

1. Configuring systems that provide name/address resolution to supply additional data origin and integrity artifacts along with the authoritative data returned in response to resolution queries; and
2. Configuring systems that provide name/address resolution to Smart Grid information systems, when operating as part of a distributed, hierarchical namespace, to provide the means to indicate the security status of child subspaces and, if the child supports secure resolution services, enabled verification of a chain of trust among parent and child domains.

Supplemental Guidance

None.

Requirement Enhancements

None.

Additional Considerations

None.

Impact Level Allocation

Low: SG.SC-21	Moderate: SG.SC-21	High: SG.SC-21

SG.SC-22 Fail in Known State

Category: Common Technical Requirements, Integrity

Requirement

The Smart Grid information system fails to a known state for defined failures.

Supplemental Guidance

Failure in a known state can be interpreted by organizations in the context of safety or security in accordance with the organization's mission/business/operational needs. Failure in a known secure state helps prevent a loss of confidentiality, integrity, or availability in the event of a failure of the Smart Grid information system or a component of the Smart Grid information system.

Requirement Enhancements

None.

Additional Considerations

A1. The Smart Grid information system preserves defined system state information in failure.

Impact Level Allocation

Low: Not Selected	Moderate: SG.SC-22	High: SG.SC-22

SG.SC-23 Thin Nodes

Category: Unique Technical Requirements

Requirement

The Smart Grid information system employs processing components that have minimal functionality and data storage.

Supplemental Guidance

The deployment of Smart Grid information system components with minimal functionality (e.g., diskless nodes and thin client technologies) reduces the number of endpoints to be secured and may reduce the exposure of information, Smart Grid information systems, and services to a successful attack.

Requirement Enhancements

None.

Additional Considerations

None.

Impact Level Allocation

Low: Not Selected	Moderate: Not Selected	High: Not Selected

SG.SC-24 Honeypots

Category: Unique Technical Requirements

Requirement

The Smart Grid information system includes components specifically designed to be the target of malicious attacks for the purpose of detecting, deflecting, analyzing, and tracking such attacks.

Supplemental Guidance

None.

Requirement Enhancements

None.

Additional Considerations

A1. The Smart Grid information system includes components that proactively seek to identify Web-based malicious code.

Impact Level Allocation

| Low: Not Selected | Moderate: Not Selected | High: Not Selected |

SG.SC-25   Operating System-Independent Applications

Category: Unique Technical Requirements

Requirement

The Smart Grid information system includes organization-defined applications that are independent of the operating system.

Supplemental Guidance

Operating system-independent applications are applications that can run on multiple operating systems. Such applications promote portability and reconstitution on different platform architectures, thus increasing the availability for critical functionality while an organization is under an attack exploiting vulnerabilities in a given operating system.

Requirement Enhancements

None.

Additional Considerations

None.

Impact Level Allocation

| Low: Not Selected | Moderate: Not Selected | High: Not Selected |

SG.SC-26   Confidentiality of Information at Rest

Category: Unique Technical Requirements

Requirement

The Smart Grid information system employs cryptographic mechanisms for all critical security parameters (e.g., cryptographic keys, passwords, security configurations) to prevent unauthorized disclosure of information at rest.

Supplemental Guidance

For a list of current FIPS-approved or allowed cryptography, *see* Chapter Four Cryptography and Key Management.

Requirement Enhancements

None.

Additional Considerations

None.

Impact Level Allocation

| Low: Not Selected | Moderate: SG.SC-26 | High: SG.SC-26 |

SG.SC-27   Heterogeneity

Category: Unique Technical Requirements

Requirement

The organization employs diverse technologies in the implementation of the Smart Grid information system.

Supplemental Guidance

Increasing the diversity of technologies within the Smart Grid information system reduces the impact from the exploitation of a specific technology.

Requirement Enhancements

None.

Additional Considerations

None.

Impact Level Allocation

| Low: Not Selected | Moderate: Not Selected | High: Not Selected |

SG.SC-28   Virtualization Techniques

Category: Unique Technical Requirements

Requirement

The organization employs virtualization techniques to present gateway components into Smart Grid information system environments as other types of components, or components with differing configurations.

Supplemental Guidance

Virtualization techniques provide organizations with the ability to disguise gateway components into Smart Grid information system environments, potentially reducing the likelihood of successful attacks without the cost of having multiple platforms.

Requirement Enhancements

None.

Additional Considerations

A1. The organization employs virtualization techniques to deploy a diversity of operating systems environments and applications;

A2. The organization changes the diversity of operating systems and applications on an organization-defined frequency; and

A3. The organization employs randomness in the implementation of the virtualization.

Impact Level Allocation

Low: Not Selected	Moderate: Not Selected	High: Not Selected

SG.SC-29 Application Partitioning

Category: Unique Technical Requirements

Requirement

The Smart Grid information system separates user functionality (including user interface services) from Smart Grid information system management functionality.

Supplemental Guidance

Smart Grid information system management functionality includes, for example, functions necessary to administer databases, network components, workstations, or servers, and typically requires privileged user access. The separation of user functionality from Smart Grid information system management functionality is either physical or logical.

Requirement Enhancements

None.

Additional Considerations

A1. The Smart Grid information system prevents the presentation of Smart Grid information system management-related functionality at an interface for general (i.e., non-privileged) users.

Additional Considerations Supplemental Guidance

The intent of this additional consideration is to ensure that administration options are not available to general users. For example, administration options are not presented until the user has appropriately established a session with administrator privileges.

Impact Level Allocation

Low: Not Selected	Moderate: Not Selected	High: SG.SC-29

SG.SC-30 Smart Grid Information System Partitioning

Category: Common Technical Requirements, Integrity

Requirement

The organization partitions the Smart Grid information system into components residing in separate physical or logical domains (or environments).

Supplemental Guidance

An organizational assessment of risk guides the partitioning of Smart Grid information system components into separate domains (or environments).

Requirement Enhancements

None.

Additional Considerations

None.

Impact Level Allocation

Low: Not Selected	Moderate: SG.SC-30	High: SG.SC-30

3.25 SMART GRID INFORMATION SYSTEM AND INFORMATION INTEGRITY (SG.SI)

Maintaining a Smart Grid information system, including information integrity, increases assurance that sensitive data have neither been modified nor deleted in an unauthorized or undetected manner. The security requirements described under the Smart Grid information system and information integrity family provide policy and procedure for identifying, reporting, and correcting Smart Grid information system flaws. Requirements exist for malicious code detection. Also provided are requirements for receiving security alerts and advisories and the verification of security functions on the Smart Grid information system. In addition, requirements within this family detect and protect against unauthorized changes to software and data; restrict data input and output; check the accuracy, completeness, and validity of data; and handle error conditions.

SG.SI-1 Smart Grid Information System and Information Integrity Policy and Procedures

Category: Common Governance, Risk, and Compliance (GRC) Requirements

Requirement

1. The organization develops, implements, reviews, and updates on an organization-defined frequency—

    a. A documented Smart Grid information system and information integrity security policy that addresses—

        i. The objectives, roles, and responsibilities for the Smart Grid information system and information integrity security program as it relates to protecting the organization's personnel and assets; and

        ii. The scope of the Smart Grid information system and information integrity security program as it applies to all of the organizational staff, contractors, and third parties; and

b. Procedures to address the implementation of the Smart Grid information system and information integrity security policy and associated Smart Grid information system and information integrity protection requirements;

2. Management commitment ensures compliance with the organization's security policy and other regulatory requirements; and

3. The organization ensures that the Smart Grid information system and information integrity policy and procedures comply with applicable federal, state, local, tribal, and territorial laws and regulations.

Supplemental Guidance

The Smart Grid information system and information integrity policy can be included as part of the general control security policy for the organization. Smart Grid information system and information integrity procedures can be developed for the security program in general and for a particular Smart Grid information system when required.

Requirement Enhancements

None.

Additional Considerations

None.

Impact Level Allocation

| Low: SG.SI-1 | Moderate: SG.SI-1 | High: SG.SI-1 |

SG.SI-2 Flaw Remediation

Category: Common Technical Requirements, Integrity

Requirement

The organization—

1. Identifies, reports, and corrects Smart Grid information system flaws;

2. Tests software updates related to flaw remediation for effectiveness and potential side effects on organizational Smart Grid information systems before installation; and

3. Incorporates flaw remediation into the organizational configuration management process.

Supplemental Guidance

The organization identifies Smart Grid information systems containing software and firmware (including operating system software) affected by recently announced flaws (and potential vulnerabilities resulting from those flaws). Flaws discovered during security assessments, continuous monitoring, or under incident response activities also need to be addressed.

Requirement Enhancements

None.

Additional Considerations

A1. The organization centrally manages the flaw remediation process. Organizations consider the risk of employing automated flaw remediation processes on a Smart Grid information system;

A2. The organization employs automated mechanisms on an organization-defined frequency and on demand to determine the state of Smart Grid information system components with regard to flaw remediation; and

A3. The organization employs automated patch management tools to facilitate flaw remediation to organization-defined Smart Grid information system components.

Impact Level Allocation

| Low: SG.SI-2 | Moderate: SG.SI-2 | High: SG.SI-2 |

SG.SI-3   Malicious Code and Spam Protection

Category: Common Governance, Risk, and Compliance (GRC) Requirements

Requirement

1. The organization—

   a. Implements malicious code protection mechanisms; and

   b. Updates malicious code protection mechanisms (including signature definitions) whenever new releases are available in accordance with organizational configuration management policy and procedures; and

2. The Smart Grid information system prevents users from circumventing malicious code protection capabilities.

Supplemental Guidance

None.

Requirement Enhancements

None.

Additional Considerations

A1. The organization centrally manages malicious code protection mechanisms;

A2. The Smart Grid information system updates malicious code protection mechanisms in accordance with organization-defined policies and procedures;

A3. The organization configures malicious code protection methods to perform periodic scans of the Smart Grid information system on an organization-defined frequency;

A4. The use of mechanisms to centrally manage malicious code protection must not degrade the operational performance of the Smart Grid information system; and

A5. The organization employs spam protection mechanisms at system entry points and at workstations, servers, or mobile computing devices on the network to detect and take action on unsolicited messages transported by electronic mail, electronic mail attachments, Web accesses, or other common means.

Impact Level Allocation

Low: SG.SI-3	Moderate: SG.SI-3	High: SG.SI-3

SG.SI-4   Smart Grid Information System Monitoring Tools and Techniques

Category: Common Governance, Risk, and Compliance (GRC) Requirements

Requirement

The organization monitors events on the Smart Grid information system to detect attacks, unauthorized activities or conditions, and non-malicious errors.

Supplemental Guidance

Smart Grid information system monitoring capability can be achieved through a variety of tools and techniques (e.g., intrusion detection systems, intrusion prevention systems, malicious code protection software, log monitoring software, network monitoring software, and network forensic analysis tools). The granularity of the information collected can be determined by the organization based on its monitoring objectives and the capability of the Smart Grid information system to support such activities.

Requirement Enhancements

None.

Additional Considerations

A1. The Smart Grid information system notifies a defined list of incident response personnel;

A2. The organization protects information obtained from intrusion monitoring tools from unauthorized access, modification, and deletion;

A3. The organization tests/exercises intrusion monitoring tools on a defined time period;

A4. The organization interconnects and configures individual intrusion detection tools into a Smart Grid system-wide intrusion detection system using common protocols;

A5. The Smart Grid information system provides a real-time alert when indications of compromise or potential compromise occur; and

A6. The Smart Grid information system prevents users from circumventing host-based intrusion detection and prevention capabilities.

Impact Level Allocation

Low: SG.SI-4	Moderate: SG.SI-4	High: SG.SI-4

SG.SI-5   Security Alerts and Advisories

Category: Common Governance, Risk, and Compliance (GRC) Requirements

Requirement

The organization—

1. Receives Smart Grid information system security alerts, advisories, and directives from external organizations; and

2. Generates and disseminates internal security alerts, advisories, and directives as deemed necessary.

Supplemental Guidance

None.

Requirement Enhancements

None.

Additional Considerations

A1. The organization employs automated mechanisms to disseminate security alert and advisory information throughout the organization.

Impact Level Allocation

| Low: SG.SI-5 | Moderate: SG.SI-5 | High: SG.SI-5 |

SG.SI-6 Security Functionality Verification

Category: Common Governance, Risk, and Compliance (GRC) Requirements

Requirement

1. The organization verifies the correct operation of security functions within the Smart Grid information system upon—

    a. Smart Grid information system startup and restart; and

    b. Command by user with appropriate privilege at an organization-defined frequency; and

2. The Smart Grid information system notifies the management authority when anomalies are discovered.

Supplemental Guidance

None.

Requirement Enhancements

None.

Additional Considerations

A1. The organization employs automated mechanisms to provide notification of failed automated security tests; and

A2. The organization employs automated mechanisms to support management of distributed security testing.

Impact Level Allocation

| Low: Not Selected | Moderate: SG.SI-6 | High: SG.SI-6 |

SG.SI-7 Software and Information Integrity

Category: Unique Technical Requirements

Requirement

The Smart Grid information system monitors and detects unauthorized changes to software and information.

Supplemental Guidance

The organization employs integrity verification techniques on the Smart Grid information system to look for evidence of information tampering, errors, and/or omissions.

Requirement Enhancements

1. The organization reassesses the integrity of software and information by performing on an organization-defined frequency integrity scans of the Smart Grid information system.

Additional Considerations

A1. The organization employs centrally managed integrity verification tools; and

A2. The organization employs automated tools that provide notification to designated individuals upon discovering discrepancies during integrity verification.

Impact Level Allocation

Low: Not Selected	Moderate: SG.SI-7 (1)	High: SG.SI-7 (1)

SG.SI-8   Information Input Validation

Category: Common Technical Requirements, Integrity

Requirement

The Smart Grid information system employs mechanisms to check information for accuracy, completeness, validity, and authenticity.

Supplemental Guidance

Rules for checking the valid syntax of Smart Grid information system input (e.g., character set, length, numerical range, acceptable values) are in place to ensure that inputs match specified definitions for format and content.

Requirement Enhancements

None.

Additional Considerations

None.

Impact Level Allocation

Low: Not Selected	Moderate: SG.SI-8	High: SG.SI-8

SG.SI-9   Error Handling

Category: Common Technical Requirements, Integrity

Requirement

The Smart Grid information system—

1. Identifies error conditions; and
2. Generates error messages that provide information necessary for corrective actions without revealing potentially harmful information that could be exploited by adversaries.

Supplemental Guidance

The extent to which the Smart Grid information system is able to identify and handle error conditions is guided by organizational policy and operational requirements.

Requirement Enhancements

None.

Additional Considerations

None.

Impact Level Allocation

| Low: SG.SI-9 | Moderate: SG.SI-9 | High: SG.SI-9 |

CHAPTER FOUR
CRYPTOGRAPHY AND KEY MANAGEMENT

This chapter identifies technical cryptographic and key management issues across the scope of systems and devices found in the Smart Grid along with potential alternatives. The identified alternatives may be existing standards, methods, or technologies, and their optimal adaptations for the Smart Grid. Where alternatives do not exist, the subgroup has identified gaps where new standards and/or technologies should be developed for the industry.

4.1 SMART GRID CRYPTOGRAPHY AND KEY MANAGEMENT ISSUES

4.1.1 General Constraining Issues

4.1.1.1 Computational Constraints

Some Smart Grid devices, particularly residential meters and in-home devices, may be limited in their computational power and/or ability to store cryptographic materials. The advent of low-cost semiconductors, including low-cost embedded processors with built-in cryptographic capabilities, will, however, ease some such constraints when the supply chain—from manufacturing to deployment to operation—absorbs this technology and aligns it with key management systems for Smart Grid operations. We can expect that most future devices connected to the Smart Grid will have basic cryptographic capabilities, including the ability to support symmetric ciphers for authentication and/or encryption. Public-key cryptography may be supported either in hardware by means of a cryptography co-processor or, as long as it is performed infrequently (i.e., less than once per hour), it can be supported in software. We also note that the use of low-cost hardware with embedded cryptography support is a necessary but not wholly sufficient step toward achieving high availability, integrity, and confidentiality in the Smart Grid. A trustworthy and unencumbered implementation of cryptography that is suitable (both computationally and resource-wise) for deployment in the Smart Grid would benefit all stakeholders in Smart Grid deployments.

4.1.1.2 Channel Bandwidth

The Smart Grid will involve communication over a variety of channels with varying bandwidths.

Encryption alone does not generally impact channel bandwidth, since symmetric ciphers such as Advanced Encryption Standard (AES) produce roughly the same number of output bits as input bits, except for rounding up to the cipher block size. However, encryption negatively influences lower layer compression algorithms, since encrypted data is uniformly random and therefore not compressible. For compression to be effective it must be performed before encryption—and this must be taken into account in designing the network stack.

Integrity protection as provided by an efficient Cipher-Based Message Authentication Code (CMAC) adds a fixed overhead to every message, typically 64 or 96 bits. On slow channels that communicate primarily short messages, this overhead can be significant. For instance, the SEL Mirrored Bits® protocol for line protection continuously exchanges 8-bit messages. Protecting these messages would markedly impact latency unless the channel bandwidth is significantly increased.

Low bandwidth channels may be too slow to exchange large certificates frequently. If the initial certificate exchange is not time critical and is used to establish a shared symmetric key or keys that are used for an extended period of time, as with the Internet Key Exchange (IKE) protocol, certificate exchange can be practical over even slow channels. However, if the certificate-based key-establishment exchange is time critical, protocols like IKE that exchange multiple messages before arriving at a pre-shared key may be too costly, even if the size of the certificate is minimal.

4.1.1.3 Connectivity

Standard Public Key Infrastructure (PKI) systems based on a peer-to-peer key establishment model where any peer may need to communicate with any other may not be necessary or desirable from a security standpoint for components in the Smart Grid. Many devices may not have connectivity to key servers, certificate authorities, Online Certificate Status Protocol (OCSP) servers, etc. Many connections between Smart Grid devices will have much longer durations (often permanent) than typical connections on the Internet.

4.1.2 General Cryptography Issues

4.1.2.1 Entropy

Many devices do not have access to sufficient sources of entropy to serve as good sources of randomness for cryptographic key generation and other cryptographic operations. This is a fundamental issue and has impacts on the key management and provisioning system that must be designed and operated in this case.

4.1.2.2 Cipher Suite

A cipher suite that is open (e.g., standards based, mature, and preferably patent free) and reasonably secure for wide application in Smart Grid systems would help enable interoperability. Factors to consider include a decision about which block ciphers (e.g., 3DES, AES) are appropriate and in which modes (CBC, CTR, etc.), the key sizes, to be used, and the asymmetric ciphers (e.g., ECC, RSA, etc.) that could form the basis for many authentication operations. The United States Federal Information Processing Standard (FIPS), the NIST Special Publications (SPs), and the NSA Suite B Cryptography strategy provide secure, standard methods for achieving interoperability. Device profile, data temporality/criticality/value should also play a role in cipher and key strength selection. FIPS 140-2 specifies requirements for validating cryptographic implementations for conformance to the FIPS and SPs.

4.1.2.3 Key Management Issues

All security protocols rely on the existence of a security association (SA). From RFC 2408, *Internet Security Association and Key Management Protocol (ISAKMP)*, "SAs contain all the information required for execution of various network security services." An SA can be authenticated or unauthenticated. The establishment of an authenticated SA requires that at least one party possess some sort of credential that can be used to provide assurance of identity or device attributes to others. In general two types of credentials are common: secret keys that are shared between entities (e.g., devices), and (digital) public key certificates for key establishment (i.e., for transporting or computing the secret keys that are to be shared). Public key certificates are used to bind user or device names to a public key through some third-party attestation model, such as a PKI.

It is not uncommon for vendors to offer solutions using secure protocols by implementing IPSec with AES and calling it a day, leaving customers to figure out how to provision all their devices with secret keys or digital certificates. It is worthwhile to ask the question, "Is it better to provision devices with secret keys or with certificates?" The provisioning of secret keys (i.e., symmetric keys) can be a very expensive process, with security vulnerabilities not present when using digital certificates. The main reason for this is that with symmetric keys, the keys need to be transported from the device where they were generated and then inserted into at least one other device; typically, a different key is required for each pair of communicating devices. Care needs to be taken to ensure that the key provisioning is coordinated so that each device receives the appropriate keys—a process that is prone to human error and subject to insider attacks. There are hardware solutions for secure key transport and loading, but these can require a great deal of operational overhead and are typically cost-prohibitive for all but the smallest systems. All of this overhead and risk can be multiplied several times if each device is to have several independent security associations, each requiring a different key. Of course, techniques like those used by Kerberos can eliminate much of the manual effort and associated cost, but Kerberos cannot provide the high-availability solution when network or power outages prevent either side of the communication link from accessing the key distribution center (KDC).

The provisioning of digital certificates can be a much more cost-effective solution, because this does not require the level of coordination posed by symmetric key provisioning. With digital certificates, each device typically only needs one certificate for key establishment, and one key establishment private key that never leaves the device, once installed. Some products generate, store, and use the private key in a FIPS-140 hardware security module (HSM). In systems like this where the private key never leaves the hardware security module, it is not hard to see how such systems can offer higher levels of security with lower associated operational costs. Of course this explanation is a bit simplistic. For example, certificate provisioning involves several steps, including the generation of a key pair with suitable entropy, the generation of a certificate signing request (CSR) that is forwarded to a Registration Authority (RA) device, appropriate vetting of the CSR by the RA, and forwarding the CSR (signed by the RA) to the Certificate Authority (CA), which issues the certificate and stores it in a repository and/or sends it back to the subject (i.e., the device authorized to use the private key). CAs need to be secured, RA operators need to be vetted, certificate revocation methods need to be maintained, certificate policies need to be defined, and so on. Operating a PKI for generating and handling certificates can also require a significant amount of overhead and is typically not appropriate for small and some midsized systems. A PKI-based solution, which can have a high cost of entry, but requires only one certificate per device (as opposed to one key per pair of communicating devices), and may be more appropriate for large systems, depending on the number of possible communicating pairs of devices. In fact, the largest users of digital certificates are the Department of Defense (DoD) and large enterprises.

4.1.2.4 Summarized Issues with PKI

A PKI is not without its issues. Most issues fall into two categories: First, a PKI can be complex to operate; and second, PKI policies are not globally understood. Both categories can be attributed to the fact that Pa KI is extremely flexible. In fact, a PKI is more of a framework than an actual solution. A PKI allows each organization to set its own policies, to define its own certificate policy Object Identifiers (CP OIDs), to determine how certificate requests are vetted, how private keys are protected, how CA hierarchies are constructed, and the allowable life of certificates and cached certificates' status information. It is exactly because of this flexibility that PKI can be expensive. Organizations that wish to deploy a PKI need to address each of these and issues, and evaluate them against their own operational requirements to determine their own specific "flavor" of PKI. Then when the organization decides to interoperate with other organizations, they need to undergo a typically expensive effort to evaluate the remote organization's PKI, compare it against the local organization's requirements, determine if either side needs to make any changes, and create an appropriate policy mapping to be used in cross-domain certificates.

Another issue affecting a PKI is the need for certificate revocation and determining the validity of a certificate before accepting it from an entity (e.g. network node) that needs to be authenticated. Typically, this is accomplished by the Relying Party (RP), the node that is performing the authentication, checking the certificate revocation list (CRL) or checking with an online certificate status server. Both of these methods typically require connectivity to a backend server. This would appear to have the same availability issues as typical server-based authentication methods, such as Kerberos- or RADIUS-based methods. However, this is not necessarily true. Methods to mitigate the reliance on infrastructure components to validate certificates are discussed under "PKI High Availability Issues [§4.1.2.4.1]."

There is also the issue of trust management. A PKI is often criticized for requiring one root CA to be trusted by everyone, but this is not actually the case. It is more common that each organization operates its own root and then cross-signs other roots (or other CAs) when they determine a need for inter-domain operations. For Smart Grid, each utility could operate their own PKI (or outsource it, if they wish). Those utility organizations that need to interoperate can cross-sign their appropriate CAs. Furthermore, it would be possible for the Smart Grid community to establish one or more bridge CAs so that utility organizations would each only have to cross-sign once with the bridge. All cross-signed certificates can and should be constrained to a specific set of applications or use cases. Trust management is not a trivial issue and is discussed in more detail under "Trust Management" [§4.1.2.4.3].

4.1.2.4.1 PKI High-Availability Issues

The seeming drawback to PKI in needing to authenticate certificates through an online server need not be seen as a major issue. Network nodes can obtain certificate status assertions periodically (when they are connected to the network) and use them at a later time when authenticating with another node. In general, with this method, the node would present its certificate status assertion along with its certificate when performing authentication; Transport Layer Security (TLS) already supports this functionality. This is commonly referred to as Online Certificate Status Protocol (OCSP) stapling. In this way, very high availability could be achieved even when the authenticating nodes are completely isolated from the rest of the network.

Symmetric key methods of establishing SAs can be classified into two general categories: server-based credentials, and preconfigured credentials. With server-based systems, such as Kerberos or RADIUS, connectivity to the security server is required for establishing a security association. Of course, these servers can be duplicated a few times to have a high level of assurance that at least one of them would always be available, but considering the size of the grid, this is not likely to offer an affordable solution that can ensure that needed SAs can always be established in the case of various system outages. Duplication of the security server also introduces unnecessary vulnerabilities. As it is impossible to ensure that every node will always have access to a security server, this type of solution may not always be suitable for high-availability use cases.

The preconfigured SA class solution requires that each device be provisioned with the credential (usually a secret key or a hash of the secret key) of every entity with whom that the device will need to authenticate. This solution, for all but the smallest systems, is likely to be excessively costly, subject to human error, and encumbered with significant vulnerabilities, due to the replication of so many credentials.

Digital certificates, on the other hand, have the distinct advantage that the first node can establish an Authenticated SA with any other node that has a trust relationship with the first node's issuing CA. This trust relationship may be direct (i.e., it is stored as a trust anchor on the second node), or it may result from a certificate chain.

In the case where a chain of certificates is needed to establish trust, it is typical for devices to carry a few types of certificates. The device would need a chain of certificates beginning with its trust anchor (TA) and ending with its own certificate. The device may also carry one or more certificate chains beginning with the TA and ending with a remote domain's TA or CA. The device can store its own recent certificate status. In a system where every node carries such data, it is possible for all "trustable" nodes to perform mutual authentication, even in the complete absence of any network infrastructure.

With using a PKI, it is important for a Relying Party (RP) to verify the status of the certificate being validated. Normally, the RP would check a CRL or verify the certificate status with an OCSP responder. Another method, proposed in RFC 4366, but not widely deployed, involves a technique called OCSP stapling. With OCSP stapling, a certificate subject obtains an OCSP response (i.e., a certificate status assertion) for its own certificate and provides it to the RP. It is typical for OCSP responses to be cached for a predetermined time, as is similarly done with CRLs. Therefore, it is possible for devices to get OCSP responses for their own certificates when in reach of network infrastructure resources and provide them to RPs at a later time. One typical strategy is for devices to attempt to obtain OCSP responses daily and cache them. Another strategy is for devices to obtain an OCSP response whenever a validation is required.

For a complete, high-assurance solution, the digital certificates must carry not only authentication credentials, but also authorization credentials. This can be accomplished in one of several ways. There are several certificate parameters that can be used to encode authorization information. Some options include Subject Distinguished Name, Extended Key Usage (EKU), the WLAN SSID extension, Certificate Policy extension, and other attributes defined in RFC 4334 and other RFCs. A complete analysis of which fields to use and how to use them would be a large undertaking suitable for its own paper on the topic. Briefly, however, it is worth mentioning that the distinguished names (DNs) option offers many subfields which could be used to indicate a type of device or a type of application that this certificate subject is authorized to communicate with. The EKU field provides an indication of protocols for which the certificate is authorized (e.g., IPSec, TLS, and Secure Shell or SSH). The WLAN SSID extension can be used to limit a device to only access listed SSIDs. The most promising extension for authorization is probably the Certificate Policy (CP) extension. The CP extension indicates to the RP the applicability of a certificate to a particular purpose.

It is also possible to encode authorization credentials into either the subject's identity certificate (which binds the subject's identity to the public key) or to encode the authorization credential into a separate attribute certificate. Typically, organizations need to weigh the benefits of needing to support only one set of certificates with the issues surrounding reissuing identity certificates every time a subject's authorization credential changes. When issuing credentials to people, this is a valid issue. For devices it is rare that authorization credentials will need to change; thus, placing the authorization credentials in the identity certificates poses few disadvantages.

With proper chains of certificates, recent OCSP responses, and authorization credentials, it is possible to provide very high assurance systems that allow two entities to authenticate for authorized services, even when significant portions of the network infrastructure are unavailable.

4.1.2.4.2 Hardware Security Module and PKI

As mentioned above, it is possible to generate and store the secret or private keys used in public key-based cryptography in an HSM. It is reasonable to ask if such devices will drive up costs for price-sensitive Smart Grid components such as sensors. Currently, the smartcard market is driving down the price of chips that can securely store keys, as well as perform public key operations. Such chips can cost only a couple of dollars when purchased in large quantities. Not only does this provide security benefits, but in addition, such chips can offload processing from the embedded device CPU during cryptographic operations. CPU processing capabilities should not then be a significant obstacle to the use of public key cryptography for new (non-legacy) devices. It is typical for public key cryptography to be required only during SA establishment. After the SA has been established, symmetric key cryptography is more favorable. However it is recognized that the supply chain (from manufacture to deployment) and asset owner operations require more Smart Grid-focused key management and encryption standards before the broad use of such technology across the entire infrastructure.

4.1.2.4.3 Trust Management

A number of high-level trust management models can be considered: strict hierarchy, full mesh, or federated trust management[22], for example. When multiple organizations are endeavoring to provide a rich web of connectivity that extends across the resources of the multiple agencies, the strict hierarchy model can quickly be eliminated, because it is typically very difficult to get everyone involved to agree on who they can all trust, and under what policies this "trusted" party should operate. Just as importantly, a strict hierarchy relies on the absolute security of the central "root of trust," because a breach of the central root destroys the security of the whole system. This leaves the mesh model and the federated trust management model. The mesh model is likely to be too expensive. In fact, the federated model brings together the best features of a hierarchy and a mesh. A PKI federation is an abstract term that is usually taken to mean a domain that controls (whether owned or outsourced) its own PKI components and policies and that decides for itself its internal structure—usually, but not always a hierarchy. The domain decides when and how to cross-sign with other domains, whether directly or through a regional bridge. Such a federated approach is really the only reasonable solution for large inter-domain systems.

---

[22] See Housley, Polk; "Planning for PKI" 2001 Chapter 10, "Building and Validating Certification Paths"

In general any two domains should be allowed to cross-sign as they see fit. However, the activity of cross-signing with many other domains can result in significant overhead. Utilitiy companies may wish to form regional consortiums that would provide bridging services for its member utility companies to help alleviate this concern.

Small utilities could outsource their PKI. This is not necessarily the same as going to a public PKI provider, such as a large CA organization, and getting an "Internet model" certificate. With the Internet model, a certificate mainly proves that you are the rightful owner of the domain name listed in your certificate. For Smart Grid, this is probably not sufficient. Certificates should be used to prove ownership, as well as being used for authorization credentials. Smart Grid certificates could be issued under Smart Grid–sanctioned policies and could carry authorization credentials.

IEEE 802.16 (WiMAX) PKI certificates, by comparison, do not prove ownership; they can only be used to prove that the entity with the corresponding private key is the entity listed in the certificate. An AAA server must then be queried to obtain the authorization credential of the device.

4.1.2.4.4 Need for a Model Policy

A certificate policy is a document that describes the policies under which a particular certificate was issued. A typical CP document contains a rich set of requirements for all PKI participants, including those that are ascribed to the Relying Party. A CP document also contains legal statements, such as liability limits that the PKI is willing to accept. RFC 3647 provides an outline and description for a template CP document. Most PKIs follow this template.

A certificate reflects the CP that it was issued under by including a Certificate Policy Extension. The CP Extension contains an Object ID that is a globally unique number string (also referred to as an arc) that can be used by an RP to trace back to a CP document. The RP can then determine information about the certificate, such as the level of assurance with which it was issued, how it was vetted, how the private keys of the CA are protected, and whether the RP should obtain recent status information about the certificate.

A CP OID also indicates the applicability of a certificate to a particular application. A PKI can use different CP OIDs for different device types to clearly distinguish between those device types, which reduces the need to rely on strict naming conventions. The RP can be configured with acceptable CP OIDs, eliminating the need for the RP to actually obtain and read the CP document.

4.1.2.4.5 Certificate Lifetimes

It is tempting to issue certificates with lifetimes of 50 years or longer. This seems convenient, because they are out there and no one needs to worry about them for 50 years. However, the use of 50-year certificates would have serious implications in the future. Revoked certificates must remain on a CRL until the certificate expires. This can create very large CRLs that are an issue for those resource-constrained devices found throughout the Smart Grid.

Certificate lifetimes should be set to an amount of time commensurate with system risks and application; however as an upper bound it is recommended a maximum of 10 years not be surpassed. An approaching expiration date should trigger a flag in the system, urging replacement of the certificate—a scheme that would reduce the burden of storing a large number of revoked certificates in the CRL.

A more appropriate solution would be to determine reasonable lifetimes for all certificates. This is not a trivial issue, and different organizations, for a variety of reasons, will select different lifetimes for similar certificates. The following points address a few considerations for three different types of certificates:

- *Manufacturers' Device Management Certificates*. These certificates are installed into devices by the manufacturer; they typically bind the make, model, and serial number of a device to a public key and are used to prove the nature of the device to a remote entity. These certificates typically offer no trust in themselves (other than to say what the device is); that is, they do not provide any authorization credentials. They can be used to determine if the device is allowed access to given resources. It is common to use this certificate to find a record in an AAA server that indicates the authorization credentials of the subject device. For such certificates, RFC 5280 (§ 4.1.2.5) recommends using a Generalized Time value of 99991231235959Z for the expiration date (i.e., the notAfter date). This indicates that the certificate has no valid expiration date.

- *User Certificates*. One of the main reasons to select a certificate lifetime is to manage the size of the associated CRLs. Factors that can affect the total number of revoked certificates in a domain include the total number of certificates issued, the certificate lifetimes, and employee turnover. Regardless of how many certificates are currently revoked, there are several other ways to manage CRL sizes. Some of these methods include partitioning the certificates across multiple CAs, scoping CRLs to portions of the user base, and implementing multiple CRL issuers per CA. The operator's Policy Management authority will have to take these considerations into account and derive their own policy. Two to three years are common lifetimes for user certificates. For example, the DoD certificate policy specifies maximum certificate lifetimes of three years for high and medium assurance certificates.

- *Operator-Issued Device Certificates*. As mentioned above for operator (e.g., utility) issued device certificates, such limitless lifetimes would not be appropriate, due to issues with maintaining CRLs. Because device turnover is typically less frequent than user turnover, it is reasonable to issue these certificates with longer lifetimes. A reasonable range to consider would be three to six years. Going much beyond six years may introduce key lifetime issues.

This is not a trivial topic, and future work should be done to ensure that appropriate guidelines and best practices are established for the Smart Grid community.

4.1.2.5 Elliptic Curve Cryptography

The National Security Agency (NSA) has initiated a Cryptographic Interoperability Strategy (CIS) for U.S. government systems. Part of this strategy has been to select a set of NIST-approved cryptographic techniques, known as NSA Suite B, and foster the adoption of these techniques through inclusion into standards of widely-used protocols, such as the Internet Engineering Task Force (IETF) TLS, Secure/Multipurpose Internet Mail Extensions (S/MIME), IPSec, and SSH. NSA Suite B consists of the following NIST-approved techniques:

- ***Encryption***. Advanced Encryption Standard – FIPS PUB 197 (with keys sizes of 128 and 256 bits)[23]
- ***Key Exchange***. The Ephemeral Unified Model and the One-Pass Diffie-Hellman key agreement schemes (two of several ECDH schemes) – NIST Special Publication 800-56A (using the curves with 256- and 384-bit prime moduli)
- ***Digital Signature***. Elliptic Curve Digital Signature Algorithm (ECDSA) – FIPS PUB 186-3 (using the curves with 256 and 384-bit prime moduli)
- ***Hashing***. Secure Hash Algorithm (SHA) – FIPS PUB 180-3 (using SHA-256 and SHA-384)

Intellectual Property issues have been cited pertaining to the adoption of ECC. To mitigate these issues NSA has stated [§4.4-25]:

> A key aspect of Suite B Cryptography is its use of elliptic curve technology instead of classic public key technology. In order to facilitate adoption of Suite B by industry, NSA has licensed the rights to 26 patents held by Certicom, Inc. covering a variety of elliptic curve technology. Under the license, NSA has the right to grant a sublicense to vendors building certain types of products or components that can be used for protecting national security information.[24]

A number of questions arise when considering this license for Smart Grid use:

1. How can vendors wishing to develop Suite B–enabled commercial off-the-shelf (COTS) products for use within the national security field obtain clarification on whether their products are licensable within the field of use?
2. What specific techniques within Suite B are covered by the Certicom license?
3. To what degree can the NSA license be applied to the Smart Grid?
4. What are the licensing terms of this technology outside the NSA sublicense?

These industry issues have produced some undesirable results:

1. Technology vendors are deploying ECC schemes based on divergent standardization efforts or proprietary specifications that frustrate interoperability.
2. Technology vendors are avoiding deployment of the standardized techniques, thwarting the adoption and availability of commercial products.
3. New standardization efforts are creating interoperability issues.

It is also worth noting that ECC implementation strategies based on the fundamental algorithms of ECC, which were published prior to the filing dates of many of the patents in this area, are identified and described in the IETF Memo entitled "Fundamental Elliptic Curve Cryptography Algorithms."[25]

---

[23] *See*, FIPS PUB 197 at the National Institute of Standards and Technology, FIPS Publications listing.

[24] *See*, http://www.nsa.gov/ia/contacts/index.shtml for more information.

[25] Available at http://tools.ietf.org/html/draft-mcgrew-fundamental-ecc-01.txt

Intellectual property rights (IPR) statements and frequently asked questions (FAQs) covering pricing have been made concerning some commercial use of patented ECC technology.[26] However, these have not been comprehensive enough to cover the envisioned scenarios that arise in the Smart Grid. Interoperability efforts, where a small set of core cryptographic techniques are standardized, as in the NSA Cryptographic Interoperability Strategy, have been highly effective in building multivendor infrastructures that span numerous standards development organizations' specifications.

Federal support and action that specifies and makes available technology for the smart energy infrastructure, similar to the Suite B support for national security, would remove many of these issues for the Smart Grid.

4.1.3 Smart Grid System-Specific Encryption and Key Management Issues – Smart Meters

Where meters contain cryptographic keys for authentication, encryption, or other cryptographic operations, a key management scheme must provide for adequate protection of cryptographic materials, as well as sufficient key diversity. That is, a meter, collector, or other power system device should not be subject to a break-once break-everywhere scenario, due to the use of one secret key or a common credential across the entire infrastructure. Each device should have unique credentials or key material such that compromise of one device does not impact other deployed devices. The key management system (KMS) must also support an appropriate lifecycle of periodic rekeying and revocation.

There are existing cases of large deployed meter bases using the same symmetric key across all meters—and even in different states. In order to share network services, adjacent utilities may even share and deploy that key information throughout both utility Advanced Metering Infrastructure (AMI) networks. Compromising a meter in one network could compromise all meters and collectors in both networks.

4.2 CRYPTOGRAPHY AND KEY MANAGEMENT SOLUTIONS AND DESIGN CONSIDERATIONS

Secure key management is essential to the effective use of cryptography in deploying a Smart Grid infrastructure. NIST SP 800-57, *Recommendation for Key Management Part 1*, recommends best practices for developers and administrators on secure key management. These recommendations are as applicable for the Smart Grid as for any other infrastructure that make use of cryptography, and they are a starting point for Smart Grid key management.[27]

4.2.1 General Design Considerations

4.2.1.1 Selection and Use of Cryptographic Techniques

Designing cryptographic algorithms and protocols that operate correctly and are free of undiscovered flaws is difficult at best. There is general agreement in the cryptographic

---

[26] *See*, http://www.certicom.com/images/pdfs/certicom%20-ipr-contribution-to-ietfsept08.pdf and http://www.certicom.com/images/pdfs/certicom%20zigbee%20smart%20energy%20faq_30_mar_2009.pdf

[27] Please see Chapter 9 R&D for a discussion of some of the considerations.

community that openly-published and time-tested cryptographic algorithms and protocols are less likely to contain security flaws than those developed in secrecy, because their publication enables scrutiny by the entire community. Historically, proprietary and secret protocols have frequently been found to contain flaws when their designs become public. For this reason, FIPS-approved and NIST-recommended cryptographic techniques are preferred, where possible. However, the unique requirements that some parts of the Smart Grid place on communication protocols and computational complexity can drive a genuine need for cryptographic techniques that are not listed among the FIPS-approved and NIST-recommended techniques. Known examples are the PE Mode as used in IEEE P1711 and EAX' as used in American National Standard (ANS) C12.22.

The general concerns are that these additional techniques have not received a level of scrutiny and analysis commensurate with the standards development process of FIPS and recommendation practices of NIST. At a minimum, a technique outside of this family of techniques should (1) be defined in a publicly available forum, (2) be provided to a community of cryptographers for review and comment for a reasonable duration, (3) be in, or under development in, a standard by a recognized standards-developing organization (SDO). In addition, a case should be made for its use along the lines of resource constraints, unique nature of an application, or new security capabilities not afforded by the FIPS-approved and NIST-recommended techniques.

4.2.1.2 Entropy

As discussed earlier in the section there are considerations when dealing with entropy on many constrained devices and systems that can be found throughout the Smart Grid. There are some possible approaches that can address restricted sources of entropy on individual point devices, they include:

- Seeding a Deterministic Random Bit Generator (DRBG) on a device before distribution; any additional entropy produced within the device could be used to reseed it.

- Alternatively, a Key Derivation Function (KDF) could derive new keys from a long-term key that the device has been pre-provisioned with.

4.2.1.3 Cryptographic Module Upgradeability

Cryptographic algorithms are implemented within cryptographic modules that need to be designed to protect the cryptographic algorithm and keys used in the system. The following need to be considered when planning the upgradeability of these modules:

- Smart Grid equipment is often required to have an average life of 20 years, which is much longer than for typical information technology (IT) and communications systems.

- Due to reliability requirements for the electrical grid, testing cycles are often longer and more rigorous.

- The replacement of deployed devices can take longer and be more costly than for many IT and communications systems (e.g., wholesale replacement of millions of smart meters).

Careful consideration in the design and planning phase of any device and system for Smart Grid needs to take the above into account.

Over time, there have been challenges with obtaining and maintaining the required level of protection when using cryptographic algorithms, protocols, and their various compositions in working systems. For example, failures in encryption systems usually occur because of one or more of the following issues ranked, in order of decreasing likelihood:

- *Implementation errors.* Examples can include poor random number generator (RNG) seeding, poor sources of entropy, erroneous coding of a protocol/algorithm, HSM application program interface (API) errors/vulnerabilities that lead to Critical Security Parameter (CSP) leakage, etc.

- *Compositional failures.* Combining cryptographic algorithms without adequate analysis, which leads to less secure systems overall.

- *Insecure protocols.* This occurs when items, such as authentication protocols, are found to be insecure while their underlying algorithms may be secure. It is a similar issue to compositional failure, but protocols are inherently more complex constructions, as they usually involve multiparty message flows and possible complex states.

- *Insecure algorithms.* The probability that basic modern cryptographic algorithms, such as symmetric/asymmetric encryption and/or hash functions would become totally insecure is relatively low, but it always remains a possibility, as new breakthroughs occur in basic number theory, cryptanalysis, and new computing technologies. What is more likely is that subtle errors, patterns, or other mathematical results that reduce the theoretical strength of an algorithm will be discovered. There is also a long term (perhaps beyond the scope of many equipment lifetimes being deployed in Smart Grid) possibility of Quantum Computing (QC) being realized. The cryptographic consequences of QC vary, but current research dictates that the most relied upon asymmetric encryption systems (e.g. RSA, ECC, DH) would fail. However, doubling key sizes for symmetric ciphers (e.g. AES 128 bit to 256 bit) should be sufficient to maintain their current security levels under currently known theoretical attacks.

When designing and planning for Smart Grid systems, there are some design considerations that can address the risks under discussion:

- The use of approved and thoroughly reviewed cryptographic algorithms is strongly advised. The NIST Computer Security Division[28] has published a wealth of such cryptographic mechanisms and implementation guidance.

- Well-understood, mature, and publicly vetted methods that have been extensively peer-reviewed by a community of cryptographers and an open standards process should be preferred over cryptographic compositions or protocols that are based on proprietary and closed development.

- Independently validated cryptographic implementations, where cost and implementation feasibility allow, should be preferred over non-reviewed or unvalidated implementations.

- Cryptographic modules (both software and hardware) that can support algorithm and key length flexibility and maintain needed performance should be preferred over those that

---

[28] *See,* http://csrc.nist.gov.

cannot be changed, in case an algorithm is found to be no longer secure or a bit-strength-reducing vulnerability is found in the cryptographic algorithm.

- Providing a cryptographic design (including, but not limited to, key length) that exceeds current security requirements in order to avoid or delay the need for later upgrade.

- Cryptographic algorithms are often used within communications protocols. To enable possible future changes to the cryptographic algorithms without disrupting ongoing operation, it is good practice to design protocols that allow alternative cryptographic algorithms. Examples can include the negotiation of security parameters, such that future changes to cryptographic algorithms may be accommodated within the protocol (e.g., future modifications, with backwards compatibility), and support the simultaneous use of two or more cryptographic algorithms during a period of transition.

- It is understood that there will be cases in which, due to cost, chip specialization to particular standards, performance requirements, or other practical considerations, a cryptographic algorithm implementation (or aspects of it, such as key length) may not be upgradeable. In such cases, it may be prudent to ensure that adequate planning is in place to treat affected devices/systems as less trusted in the infrastructure and, for example, use enhanced network segmentation, monitoring, and containment (upon possible intrusion or tampering detection).

4.2.1.4 Random Number Generation

Random numbers or pseudorandom numbers are frequently needed when using cryptographic algorithms, e.g., for the generation of keys and challenge/responses in protocols. The failure of an underlying random number generator can lead to the compromise of the cryptographic algorithm or protocol and, therefore, the device or system in which the weakness appears.

Many Smart Grid devices may have limited sources of entropy that can serve as good sources of true randomness. The design of a secure random number generator from limited entropy is notoriously difficult. Therefore, the use of a well-designed, securely seeded and implemented deterministic random bit generator (i.e., also known as a pseudorandom number generator) is required. In some cases Smart Grid devices may need to include additional hardware to provide a good source of true random bits for seeding such generators.

There are several authoritative sources of information on algorithms to generate random numbers. One is NIST SP 800-90, *Recommendation for Random Number Generation Using Deterministic Random Bit Generators (Revised)*. [§4.4-18]

Another source is the multi-part American National Standard (ANS) X9.82 Standard being developed within ASC X9. Part 1 is "Overview and Basic Principles," Part 2 is "Entropy Sources," Part 3 is "Deterministic Random Bit Generators (DRBGs)," and Part 4 is "Random Bit Generation Constructions." As of February 2010, only Parts 1 and 3, published in 2006 and 2007, respectively, are available as published standards. Note that Part 3 of ANS X9.82 contains three of the four DRBGs contained within NIST SP 800-90.

NIST and ANSI have been collaborating and continue to collaborate closely on this work.

NIST has also published NIST SP 800-22, *A Statistical Test Suite for Random and Pseudorandom Number Generators for Cryptographic Applications* [§4.4-11], which provides a comprehensive description of a battery of tests for RNGs that purport to provide non-biased

output. Both the report and the software may be obtained from
http://csrc.nist.gov/groups/ST/toolkit/rng/documentation_software.html.

4.2.1.5 Local Autonomy of Operation

It may be important to support cryptographic operations, such as authentication and authorization, when connectivity to other systems is impaired or unavailable. For example, during an outage, utility technicians may need to authenticate to devices in substations to restore power, and must be able to do so even if connectivity to the control center is unavailable. Authentication and authorization services must be able to operate in a locally autonomous manner at the substation.

4.2.1.6 Availability

Availability for some (but not all) Smart Grid systems can be more important than security. Dropping or refusing to re-establish connections due to key or certificate expiration may interrupt critical communications.

If one endpoint of a secure communication is determined by a third party to have been compromised, it may be preferable to simply find a way of informing the other endpoint. This is true whether the key management is PKI or symmetric key-based. In a multi-vendor environment, it may be most practical to use PKI-based mechanisms to permit the bypass or deauthorization of compromised devices (e.g., by revocation of the certificates of the compromised devices).

4.2.1.7 Algorithms and Key Lengths

NIST SP 800-57, *Recommendation for Key Management* [§4.4-15] recommends the cryptographic algorithms and key lengths to be used to attain given security strengths. Any KMS used in the Smart Grid should carefully consider these guidelines and provide rationale when deviating from these recommendations.

4.2.1.8 Physical Security Environment

The protection of Critical Security Parameters (CSPs), such as keying material and authentication data, is necessary to maintain the security provided by cryptography. To protect against unauthorized access, modification, or substitution of this data, as well as device tampering, cryptographic modules can include features that provide physical security.

There are multiple embodiments of cryptographic modules that may provide physical security, including: multichip standalone, multichip embedded, and single-chip devices. Specific examples of such device types providing cryptographic services and physical security include Tamper Resistant Security Modules (TRSMs), Hardware Security Modules, Security Authentication Module cards (SAM cards), which may have been validated as FIPS 140-2 cryptographic modules.

Physical protection is an important aspect of a module's ability to protect itself from unauthorized access to CSPs and tampering. A cryptographic module implemented in software and running on an unprotected system, such as a general-purpose computer, commonly does not have the ability to protect itself from physical attack. When discussing cryptographic modules, the term "firmware" is commonly used to denote the fixed, small, programs that internally control a module. Such modules are commonly designed to include a range of physical security protections and levels.

In determining the appropriate level of physical protections required for a device, it is important to consider both the operating environment and the value and sensitivity of the data protected by the device. Therefore, the specification of cryptographic module physical protections is a management task in which both environmental hazard and data value are taken into consideration. For example, management may conclude that a module protecting low value information and deployed in an environment with physical protections and controls, such as equipment cages, locks, cameras, and security guards, etc., requires no additional physical protections and may be implemented in software executing on a general purpose computer system. However, in the same environment, cryptographic modules protecting high value or sensitive information, such as root keys, may require strong physical security.

In unprotected or lightly protected environments, it is common to deploy cryptographic modules with some form of physical security. Even at the consumer level, devices that process and contain valuable or sensitive personal information often include physical protection. Cable Television Set-top boxes, DVD players, gaming consoles, and smart cards are examples of consumer devices. Smart Grid equipment, such as smart meters, deployed in similar environments will, in some cases, process information and provide functionality that can be considered sensitive or valuable. In such cases, management responsible for meter functionality and security may determine that meters must include cryptographic modules with a level of physical protection.

In summary, cryptographic modules may be implemented in a range of physical forms, as well as in software on a general purpose computer. When deploying Smart Grid equipment employing cryptographic modules, the environment, the value of the information, and the functionality protected by the module should be considered when assessing the level of module physical security required.

4.2.2 Key Management Systems for Smart Grid

4.2.2.1 Public Key Infrastructure

4.2.2.1.1 Background

Certificates are issued with a validity period. The validity period is defined in the X509 certificate with two fields called "notBefore" and "notAfter." The notAfter field is often referred to as the expiration date of the certificate. As will be shown below, it is important to consider certificates as valid only if they are being used during the validity period.

If it is determined that a certificate has been issued to an entity that is no longer trustworthy (for example the certification was issued to a device that was lost, stolen, or sent to a repair depot), the certificate can be revoked. Certificate revocation lists are used to store the certificate serial number and revocation date for all revoked certificates. An entity that bases its actions on the information in a certificate is called a Relying Party (RP). To determine if the RP can accept the certificate, the RP needs to check the following criteria, at a minimum:

1. The certificate was issued by a trusted CA. (This may require the device to provide or the RP to obtain a chain of certificates back to the RP's trust anchor.)

2. The certificates being validated (including any necessary chain back to the RP's trust anchor) are being used between the notBefore and notAfter dates.
3. The certificates are not in an authoritative CRL.
4. Other steps may be required, depending on the RP's local policy, such as verifying that the distinguished name of the certificate subject or the certificate policy fields are appropriate for the given application for which the certificate is being used.

This section focuses primarily on steps 2 and 3.

4.2.2.1.2 Proper Use of Certificate Revocation, and Expiration Dates of Certificates

As mentioned above, when a certificate subject (person or device) is no longer trustworthy or the private key has been compromised, the certificate is placed into a CRL. This allows RPs to check the CRL to determine a certificate's validity status by obtaining a recent copy of the CRL and determining whether or not the certificate is listed. Over time, a CRL can become very large as more and more certificates are added to the revocation list, (e.g., devices are replaced and no longer needed, but the certificate has not expired). To prevent the CRL from growing too large, PKI administrators determine an appropriate length of time for the validity period of the certificates being issued. When a previously revoked certificate has expired, it need no longer be kept on the CRL, because an RP will see that the certificate has expired and would not need to further check the CRL.

Administrators must consider the balance between issuing certificates with short validity periods and more operational overhead, but with more manageably-sized CRLs, against issuing certificates with longer validity periods and lower operational overhead, but with potentially large and unwieldy CRLs.

When certificates are issued to employees whose employment status or level of responsibility may change every few years, it would be appropriate to issue certificates with relatively short lifetimes, such as a year or two. In this way, if an employee's status changes and it becomes necessary to revoke his/her certificate, then this certificate would only need to be maintained on the CRL until the certificate expiration date. In this way (by issuing relatively short life certificates), the CRLs can be kept to a reasonable size.

When certificates are issued to devices that are expected to last for many years, and these devices are housed in a secure environment, it may not be necessary to issue a certificate with such short validity periods, as the likelihood of ever needing to revoke a certificate is low. Therefore, the CRLs would not be expected to be very large. The natural question then arises: When a Smart Grid RP receives a certificate from an entity (person or device), and the certificate has expired, should the RP accept the certificate and authenticate the entity, or should the RP reject the certificate? What if rejecting the certificate will cause a major system malfunction?

First, consider that Smart Grid devices will be deployed with the intent to keep them operational for many years (probably in the neighborhood of 10 to 15 years). Therefore, replacing these devices should not occur very often. Of course, there will be unplanned defects that will cause devices to be replaced from time to time. The certificates of these defective devices will need to be listed on the CRL when the devices are removed from service, unless their keys can be guaranteed to be securely destroyed. In order to avoid the unlimited growth of CRLs, it would be prudent to issue device certificates with an appropriate lifetime. For devices expected to last 20 years, which are housed in secure facilities, and have a low mean-time-before-failure (MTBF), a 10-year certificate may be appropriate. This means that when a device having a certificate of this length is installed in the system and subsequently fails, it may need to be on a CRL for up to ten years.

If a good device never gets a new certificate before its certificate expires, the device will no longer be able to communicate in the system. To avoid this, the device could be provisioned with a "renewed" certificate quite some time before its current certificate expires. For example, the device may be provisioned with a new certificate a year before its current certificate expires. If the renewal attempt failed for any reason, the device would have a whole year to retry to obtain a new certificate. It is therefore easy to see that the probability of a critical device not being able to participate in the system because of an expired certificate can be made as low as desirable by provisioning the device with a new certificate sufficiently before the expiration of the old certificate."

It is worth mentioning that because of the size and scale of the Smart Grid, other techniques may be needed to keep CRLs from growing excessively. These would include the partitioning of CRLs into a number of smaller CRLs by "scoping" CRLs, based on specific parameters, such as the devices' location in the network, the type of device, or the year in which the certificate was issued. Methods for supporting such partitioning are documented in RFC 5280. Clearly with a system as large as the Smart Grid, multiple methods of limiting the size of CRLS will be required, but only with the use of reasonable expiration dates can CRLs be kept from growing without limit.

These methods should not be confused with techniques such as Delta CRLs, which allow CRLs to be fragmented into multiple files; or the use of OCSP, which allows an RP or certificate subject to obtain the certificate status for a single certificate from a certificate status server. These methods are useful for facilitating the efficient use of bandwidth; however they do nothing to keep the size of the CRLs reasonable.

4.2.2.1.3 High Availability and Interoperability Issues of Certificates and CRLs

Certificate-based authentication offers enormous benefits regarding high availability and interoperability. With certificate-based authentication, two entities that have never been configured to recognize or trust each other can "meet" and determine if the other is authorized to access local resources or participate in the network. Through a technique called "cross-signing" or "bridging" these two entities may even come from different organizations, such as neighboring utilities, or a utility and a public safety organization. However, if CRLs are stored in central repositories and are not reachable by RPs from time to time, due to network outages, it would not always be possible for RPs to determine the certificate status of the certificates that it is validating. This problem can be mitigated in a number of ways. CRLs can be cached and used by RPs for lengthy periods of time, depending on local policy. CRLs can be scoped to small geographically-close entities, such as all devices in a substation and all entities that the substation may need to communicate with. These CRLs can then be stored in the substation to enhance their accessibility to all devices in the substation. One other alternative, which has the potential of offering very high availability, is where each certificate subject periodically obtains its own signed certificate status and carries it with itself. When authenticating with an RP, the certificate subject not only provides its certificate, but also provides its most recent certificate status. If no other status source is available to the RP, and if the provided status is recent enough, the RP may accept this status as valid. This technique, sometimes referred to as OCSP stapling, is supported by the common TLS protocol and is defined in RFC 4366. OCSP stapling offers a powerful, high-availability solution for determining a certificate's status.

4.2.2.1.4 Other Issues Relating to Certificate Status

A number of additional considerations with respect to certificate status issues are as follows:

- Smart Grid components may have certificates issued by their manufacturer. These certificates would indicate the manufacturer, model and serial number of the device. If so, Smart Grid operators (e.g., utility companies) should additionally issue certificates containing specific parameters indicating how the device is being used in the system. For example, certificate parameters could indicate that the subject (i.e., the device) is owned by Utility Company X, it is installed in Substation Y, and is authorized to participate in Application Z. These certificates could be new identity certificates that also contain these new attributes (possibly in the form of Certificate Policy extensions) or they may be separate attribute certificates. Both options should be considered. For certificates issued to humans, attribute certificates may offer a more flexible solution, since human roles change. For certificates issued to devices, identity certificates that include attributes may offer a lower cost solution.

- Standardized Trust Management mechanisms would include cross-signing procedures, policy constraints for cross-signed certificates, requirements for local and regional bridge providers, as well as approved methods for issuing temporary credentials to entities during incidents involving exceptional system outages. Ideally, such methods for issuing temporary credentials would not be needed, as all entities would have their proper credentials before such an incident occurred. However, it is not unusual after a large scale incident, such as a hurricane, earthquake, or a terrorist attack, that resources would be sent across the country from sources that were never anticipated. There seem to be two general categories of solutions for such incidents. One is to make sure that all possible parties trust each other beforehand. This type of solution may require too much risk, far too much operational overhead, and unprecedented (and probably unnecessary) levels of trust and cooperation. The other method is to have a means of quickly issuing temporary local credentials to resources that arrive from remote sources. This method might rely on the resource's existing credentials from a remote domain to support the issuance of new local credentials, possibly in the form of an attribute certificate.

- Standardized certificate policies for the Smart Grid would aid interoperability. Similar standards have been successful in other industries, such as health care (ASTM standard E2212-02a, "Standard Practice for Healthcare Certificate Policy"). At one extreme, this standard set of policies would define all possible roles for certificate subjects, all categories of devices, and specific requirements on the PKI participants for each supported assurance level. Furthermore, such standards could include accreditation criteria for Smart Grid PKI service providers.

- Additional thought needs to go into determining what should be authenticated between Smart Grid components. One could argue that not only is the identity of a component important, but also its authorization status and its tamper status. The authorization status can be determined by roles, policies, or other attributes included in a certificate.

However, to determine a device's tamper status, the device will need to incorporate methods, such as high assurance boot, secure software management, and local tamper detection via FIPS 140 mechanisms. Furthermore, the device will need to use remote device attestation techniques to prove to others that it has not been tampered with.

- Some certificate subjects (i.e., devices or people) should have secure hardware for storing private keys and trust anchor certificates. Due to the advent of the Smart Card market, such mechanisms have become very affordable.

- RPs should have access to a reasonably accurate, trustworthy time source to determine if a certificate is being used within its validity period.

- Further consideration should go into determining appropriate certificate lifetimes.

4.2.2.1.5 Certificate Revocation List Alternatives

There are two alternatives to a full-blown CRL; they are CRL partitions and OCSP. A CRL partition is simply a subset of a CRL; implementations exist that have partition tables with the status of as few as 100 certificates listed in it. For example, if a device needs to validate certificate number 3456, it would send a partition request to the domain CA, and the CA would send back a partition that addresses certificates 3400–3499. The device can use it to validate if the partner (or any other certificate in that range) has been revoked. Seeing that infrastructures are typically fixed, it is probable that a device will only interact with 1–20 other devices over its entire lifetime. So requesting and storing 20 ~1 kb partition files is feasible, compared to requesting and storing an "infinitely long" CRL.

The other alternative is the Online Certificate Status Protocol, which as the name implies, is an online, real-time service. OCSP is optimal in its space requirements, as the OCSP server only stores valid certificates; there is no issue of an infinitely long CRL; the OCSP repository is only as long as the number of valid certificates in the domain. Also OCSP has the added benefit of a real-time, positive validation of a certificate. With OCSP, when a device needs to validate a potential partner, it simply sends a validation request to OCSP Responder, which simply sends back an "OK" or "BAD" indication. This approach requires no storage on the fielded device, but it does require the communications link to be active.

4.2.2.1.6 Trust Roots

A typical Web browser ships with a large number of built-in certificates (e.g., some modern browsers with up to 140). It may not be appropriate for all of the Certificate Authorities that issue these certificates to be trust roots for Smart Grid systems. On the other hand, with third-party data services and load management services, it may not be appropriate for the utility company to be the sole root of trust.

Additionally, there is a question about who issues certificates and how the system can assure that the claimed identity actually is the certificate subject. The common method for Internet use is that there are top-level (root) certificates that are the basis of all trust. This trust may be extended to secondary certificate-issuing organizations, but there is a question about how a root organization becomes a root organization, how they verify the identity for those requiring certificates, and even what identity actually means for a device.

4.2.2.2 Single Sign On

Many Smart Grid components, such as wireless devices (e.g., AMI), are low-processing-power devices with wireless interface (e.g., Zigbee) and are often connected to the backhaul networks with low bandwidth links. These components are typically equipped with 4–12 kb of RAM and 64–256 kb of flash memory. The link characteristics can also vary, depending upon the wireless radio features, such as the sleeping or idle mode of operation. For example, the advanced metering system may periodically be awakened and synced with the network to save power, rather than remain always active. Additional device requirements include (1) the support of multi-hop networks using mesh topology (e.g., to extend the backhaul reach back), and (2) support of multiple link layer technologies.

Advanced meters can also be used for other purposes besides simple metering data. For example, ANS C12.22 [§4.4-21] allows using advanced meters peering via relay or concentrators. Other applications should be able to run simultaneously on a single meter. For security requirements, each application needs to be authenticated and needs to preserve the integrity of the data provided to the system (e.g., billing system). In such scenarios, the protocol overhead and performance must be optimized, and performance must be taken into account for these low-processing power components.

From a key management perspective, optimization on the amount of exchanges and the footprint to execute peer authentication, key establishment, key update, and key deletion have to be considered for each communication layer and protocol that is used by Smart Grid components that need to be secured. This can be achieved by introducing the notion of single sign-on (SSO) to Smart Grid components (e.g., smart meters) so that one execution of peer authentication between a Smart Grid component and an authentication server can generate keys for multiple protocols within the same communication layer or across multiple communication layers. In a typical use case scenario, a smart meter may perform network access authentication based on public-key cryptography that generates a root key from which encryption keys are derived to protect each application, as well as the link-layer connection. The advantage of this scheme is that the computationally intensive public-key operation is required only once to generate the root key.

For example, the Extensible Authentication Protocol (EAP) [§4.4-22] supports multiple authentication methods called EAP methods, and its key management framework [§4.4-23] defines a key hierarchy for the Extended Master Session Key (EMSK), from which Usage-Specific Root Keys (USRKs) are derived to bootstrap encryption keys for multiple usages [§4.4-24]. EAP therefore can be a basis of SSO for smart meters. RFC 5295 [§4.4-24] also defines the key naming rule for USRK.

4.2.2.3 Symmetric Key Management

Symmetric key environments—often referred to as secret key—use a single key to both apply cryptographic protection to data (e.g., encrypt) and process cryptographically protected data (e.g., decrypt). Thus, a single key must be shared between two or more entities that need to communicate. As with any cryptographic system, there are advantages and disadvantages to this type of system. Symmetric cipher systems, relative to asymmetric ciphers, handle large amounts of data more efficiently. Symmetric keys often have a shorter lifespan than asymmetric keys, because of the amount of data that is protected using a single key; limiting the amount of data that is protected by a symmetric key helps reduce the risk of compromise of both the key and thee data. This poses important challenges in the management of these keys. The primary considerations encompassing symmetric key management includes key generation, key distribution, and key agility (i.e., the ability to change keys quickly when needed to protect different data).

The protection of the symmetric key is paramount in this type of system and is the greatest challenge in symmetric key system management. The generation of a symmetric key can essentially be accomplished in two ways: (1) locally, on the end device platform, or (2) remotely, at a single facility not physically attached to the end device platform. In the local generation scenario, a Diffie-Hellman key agreement process provides a good example for this style of generation. A simplistic description of Diffie-Hellman involves two parties that use private information known by each party and public information known by both parties to compute a symmetric key shared between the two parties. In this case, no outside influences are involved in key generation, only information known by the parties that wish to communicate is used. However, local key generation is not always possible, due to end device limitations, such as limited processor power and local memory constraints for storage of the values needed for computation.

In the remote generation scenario, the symmetric key is generated by one entity (e.g., a key server) and transported to one or more other entities (e.g. the end points that will use the key—the key consumer's device). Placement of the symmetric key into the end points can be accomplished using multiple methods that include preplaced keys or electronically distributed keys. In the preplaced method, the symmetric key is manually entered (i.e., physically loaded) into the key consuming device prior to the use of the key. This can be achieved at the factory or done when the device is deployed into the field. Electronically distributed keys need to be protected as they transit across the network to their destination. This can be achieved by encrypting the symmetric key so that only the end device can decrypt the key.

The remote generation scenario has more complexity associated with it because of distribution and trust risks. However, in the remote generation and distribution model, the concept of Perfect Forward Secrecy (PFS) can be managed for a large population of devices. PFS is dependent on the use of an ephemeral key, such that no previously used key is reused. In remote or central key generation and distribution models, PFS can be ensured because the key generation node can keep track of all previously used keys.

The preparation of the symmetric keys to be used needs to take into account both the organization (i.e., crypto groups) of which devices receive a given symmetric key and the set of keys for those devices that are needed to provide key agility. Thus, organizational management of symmetric key groups is critical to retaining control of the symmetric key as it is distributed.

Another area for consideration relative to physical key distribution is the method to establish the trust relationship between the end device and a key loader[29]—a topic beyond the scope of this section, but mentioned here for the sake of completeness. In actual practice, it will be necessary for the system managers to determine how this trust relationship is established. Establishing the

---

[29] A key loader is a device that is used to load keys directly into a device that performs encryption operations. A usage example would be in cases s where connectivity to the encryption platform has been lost and field personnel need to physically transport the keys to the encryption platform.

trust relationship should be based on a number of factors that focus on risks to the physical transport of the keys to the end point.

In the electronic distribution scenario where the symmetric key is generated by a key server that is external to the key consumer (i.e., the end point), the trust problem and the protection of the symmetric key in transit are paramount considerations to the successful implementation of this scenario. To mitigate the risk of disclosure, the key should be transported to the key consumer by wrapping (i.e., encrypting) the plaintext symmetric key, used for data protection, with a key encryption key (KEK). An individual KEK can be created by using the public key issued to the key consumer device. This way the symmetric key can be wrapped by the key generation server using the end devices public key and only unwrapped by the end devices private key. By using this method only the key consumer is able to extract the symmetric key, because only the key consumer has the associated private key, which of course remains protected on the key consumer's platform.

As can be seen, in symmetric key systems that distribute the operational key via an electronic method, a high level of coordination must be accomplished between the key producer and the key consumers. This means that a large amount of coordination management is levied on the key producer. Some considerations that the key producer must take into account include knowing exactly what group of key consumers receive the same symmetric key, risks to the key distribution channel, the key schedule to ensure that the key consumer has the right key at the right time, and how to recover from a key compromise. There are distinct advantages to remote key generation, especially since many of the devices in the Smart Grid may have limited resources, such as the processor power needed for key generation, physical memory to hold the algorithms to locally generate the symmetric key (e.g., random number generators), and the associated communications overhead to ensure that the proper key is used between the end points.

The final topic to discuss in symmetric key management is that of key agility. Key agility becomes critical when a compromise takes place as well as in normal operational mode and is directly related to preparation of the symmetric keys for use. In the case of a key compromise, key agility allows the key consumer to change to another key so that uninterrupted communication between end points can continue. However, key agility must be part of the overall key management function of planning and distribution. The key distribution package must also contain enough key material to provide operational keys plus have key material to support a compromise recovery. In the scenario where a compromise takes place, the compromise recovery key would be used, which would allow the key distribution point enough time to generate a new key package for distribution. Additionally the compromise recovery key may not be part of the same numerical branch as the previously used key to prevent a follow-on compromise where the attacker was able to determine the roll over key, based on the previously compromised key.

In the normal operational scenario where the key's lifetime comes to a natural end, the next key needs to be available to all key consumers within the same crypto group[30] prior to usage in order to ensure continuous communications. It should be noted that key roll over and the roll over

---

[30] A crypto group is a group of end devices that share a common symmetric key thereby creating a cryptographic group.

strategy is highly dependent on how the system uses the symmetric key and the frequency of communications using that key. Thus, in a scenario where communications is infrequent and the key distribution channel is secure, only a single key might be distributed to the consumer devices.

The ultimate decision on how to manage the symmetric key environment must rely on a risk assessment that considers such factors as key consumption frequency, the amount of data to be processed by the key, the security and capacity of the distribution channel, the number of symmetric keys required, and the methodology used to distribute the symmetric keys.

4.3 NISTIR HIGH-LEVEL REQUIREMENT MAPPINGS

4.3.1 Introduction

There is a need to specify cryptographic requirements and key management methods to be used in security protocols and systems that can fulfill the high-level CIA requirements. The source material that will be used to build these cryptographic requirements is in [§4.4-3] and [§4.4-4]. In summary, the high-level requirements (HLR) define low, moderate, and high levels for confidentiality, integrity, and availability, and each of these CIA requirements are mapped against the current 22 interface categories.

The interface categories are meant to capture the unique function and performance aspects of the classes of systems and devices in the Smart Grid. The cryptographic requirements that will be recommended, including those for key management, take into account the performance, reliability, computation, and communications attributes of systems and devices found in each interface category. In other words, best efforts where made to make sure that whatever is recommended should be technically and economically feasible and appropriate to the risk that must be addressed. The requirements mapping will be based on a framework for KMS attributes whose properties can be quantitatively and qualitatively analyzed for their application to the high-level requirements. Specifically, KMS attributes will be matched against the low, moderate, and high CIA levels. They will be the same for both Confidentiality and Integrity, since the capabilities and qualities of the KMS should default to the higher-level requirement in the case of cryptography. In terms of specific cryptographic suites of algorithms and key lengths, the cryptographic period requirements of NIST SP 800-57 should be used, as these requirements are not governed by anything to be found in the HLR, but by the intended lifetime of systems and their data or communication messages.

The framework of the mapping will consist of an identified cryptographic suite that is NIST-approved (i.e., FIPS-approved and/or NIST recommended) or allowed, as well as a KMS requirements matrix that maps to the HLR definitions of low, moderate, and high. The KMS matrix is a base-line for all the interface categories and can be adjusted for specific interface categories to take specific technical and risk based reasoning into account.

4.3.2 Framework

4.3.2.1 NIST-Approved Cipher Suite for Use in the Smart Grid

4.3.2.1.1 Introduction

Because Smart Grid devices can have a long operating life, the selection of cryptographic algorithms, key length, and key management methods should take into consideration the NIST transition dates specified in the following. This document lists all of the FIPS 140-2 Approved and allowed Security Functions, Random Number Generators, and Key Establishment Techniques as identified in FIPS 140-2 Annexes A, C, and D (as of 5/11/2010) and identifies which of these will be phased out by NIST as indicated in the following NIST documents:

- SP 800-57 [§4.4-15]

- SP 800-131, *DRAFT Recommendation for the Transitioning of Cryptographic Algorithms and Key Sizes* [§4.4-20]

It is important to note that the information provided in this document (i.e., NISTIR 7628) is based on the following:

- SP 800-131 is in Draft form. It is accounted for in this document because of the algorithm transition changes between 2011 and 2015. This document will be updated when the final version of SP 800-131 is released.

- The algorithms/key lengths in this document are relevant and important for NEW Implementations and those that will last beyond the year 2015. For existing implementations (i.e., validated FIPS modules), there is an expected "transition period that is provided in SP 800-131.

4.3.2.1.2 Background

All of the cryptographic algorithms that are required for use in the Smart Grid shall be NIST-approved as they currently exist today and as referenced in this report. During the development of updated versions of this report, a liaison shall be appointed to coordinate with NIST's Cryptographic Technology Group to ensure that any new algorithms are NIST-approved or allowed, and not scheduled to be withdrawn.

4.3.2.1.3 Rationale

The CSWG is chartered to coordinate cyber security standards for the Smart Grid. Since one of the primary goals is interoperability, the CSWG needs to ensure that any standards under consideration be usable by all stakeholders of the Smart Grid.

In the area of cryptography, federal law[31] requires that U.S. federal government entities must use NIST-approved or allowed algorithms. From FIPS-140-2: [§4.4-5]

> 7. Applicability. This standard is applicable to all Federal agencies that use cryptographic-based security systems to protect sensitive information in computer and telecommunication systems (including voice systems) as defined in Section 5131 of the

---

[31] The Federal Information Security Management Act of 2002; the Information Technology Management Reform Act of 1996

> Information Technology Management Reform Act of 1996, Public Law 104-106. This standard shall be used in designing and implementing cryptographic modules that Federal departments and agencies operate or are operated for them under contract. Cryptographic modules that have been approved for classified use may be used in lieu of modules that have been validated against this standard. The adoption and use of this standard is available to private and commercial organizations.

Given that many participants in the Smart Grid (including AMI) are U.S. federal agencies, interoperability requires that CSWG-listed standards be usable by them. Examples are the Tennessee Valley Authority, Bonneville Power Administration, and military bases around the world.[32]

Finally, a team of NIST cryptographers and the broader cryptographic community and general public, under a rigorous process, have reviewed the NIST-approved or allowed cryptographic suite. The goal of this robust process is to identify known weaknesses.

Examination of exceptions to the requirement:

The CSWG understands that there may exist standards and systems that take exception to this position on sound technical grounds and are potentially equally secure. The CSWG will consider these alternatives, based on submitted technical analysis that explains why the existing NIST-approved or allowed cryptographic suite could not be used. If the CSWG believes that the submitted technical analysis is sound, the CSWG will submit these other algorithms, modes, or any relevant cryptographic algorithms to NIST to be evaluated for approval for use in Smart Grid systems.

---

[32] A list of DOE-specific entities may be found at http://www.energy.gov/organization/powermarketingadmin.htm and http://www.energy.gov/organization/labs-techcenters.htm.

FIPS 140-2 Annex A: Approved Algorithms

Table 4-1 Symmetric Key – Approved Algorithms

Name	Algorithms/Key Lengths for use between 2011-2029 (per SP 800-57 and SP 800-131)	Algorithms/Key Lengths for use now and beyond 2030 (per SP 800-57 and SP 800-131)	References
Advanced Encryption Standard (AES)	All algorithms/key lengths listed in the next column are Approved during this time.	AES-128, AES-192, and AES-256 with ECB, CBC, OFB, CFB-1, CFB-8, CFB-128, CTR, or XTS mode.	National Institute of Standards and Technology, *Advanced Encryption Standard (AES)*, Federal Information Processing Standards Publication 197, November 26, 2001. National Institute of Standards and Technology, *Recommendation for Block Cipher Modes of Operation. Methods and Techniques*, Special Publication 800-38A, December 2001. National Institute of Standards and Technology, *Recommendation for Block Cipher Modes of Operation: The XTS-AES Mode for Confidentiality on Storage Devices*, Special Publication 800-38E, January 2010.
Triple-Data Encryption Algorithm (TDEA) or Triple-Data-Encryption-Standard (Triple-DES or TDES)	3-key TDES with TECB, TCBC, TCFB, TOFB, or CTR mode. (Note: 2-key TDES has 80 bits of security strength. All new implementations should have 112 bits of security strength or higher.)	N/A – cannot use TDES beyond 2030 (Note: 2-key TDES and 3-key TDES are not Approved because they have <128 bits of security.)	National Institute of Standards and Technology, *Recommendation for the Triple Data Encryption Algorithm (TDEA) Block Cipher*, Special Publication 800-67, May 2004. National Institute of Standards and Technology, *Recommendation for Block Cipher Modes of Operation. Methods and Techniques*, Special Publication 800-38A, December 2001. Appendix E references Modes of Triple-DES.

Table 4-2 Asymmetric Key – Approved Algorithms

Name	Algorithms/Key Lengths for use between 2011-2029 (per SP 800-57 and SP 800-131)	Algorithms/Key Lengths for use now and beyond 2030 (per SP 800-57 and SP 800-131)	References
Digital Signature Standard (DSS):	DSA with (L=2048, N=224) or (L=2048, N=256)	DSA with (L=3072, N=256)**	National Institute of Standards and Technology, *Digital Signature Standard (DSS)*, Federal Information Processing Standards Publication 186-3, June, 2009. (DSA, RSA2 and ECDSA2)
Digital Signature Algorithm (DSA)	RSA with (|n|=2048)	RSA with (|n|=3072)**	National Institute of Standards and Technology, *Digital Signature Standard (DSS)*, Federal Information Processing Standards Publication 186-2, January, 2000 with Change Notice 1. (DSA, RSA and ECDSA)
RSA digital signature algorithm (RSA)	ECDSA2 with curves P-224, K-233, or B-233	ECDSA2 with curves P-256, P-384, P-521, K-283, K-409, K-571, B-283, B-409, B-571	RSA Laboratories, *PKCS#1 v2.1: RSA Cryptography Standard*, June 14, 2002.
Elliptic Curve Digital Signature Algorithm (ECDSA)	Additionally, all algorithms/key lengths listed in the next column are Approved during this time. (Note: FIPS 186-2 algorithms should not be used because they are being phased out by NIST. DSA with (L=1024, N=160), RSA with (|n|=1024), and ECDSA curves K-163, B-163, P-192 have <112 bits of security.) All new implementations should have 112 bits of security strength or higher.)	**FIPS 186-3 recommends that the use of DSA with (L=3072, N=256) and RSA with (|n|=3072) should be limited to Certificate Authorities (CAs) (FIPS 186-3, Sections 4.2 and 5.1). (Note: FIPS 186-2 algorithms should not be used because they are being phased out by NIST. Key sizes less than those listed above are not Approved because they have <128 bits of security.)	Only the versions of the algorithms RSASSA-PKCS1-v1_5 and RSASSA-PSS contained within this document shall be used.

Table 4-3 Secure Hash Standard (SHS) – Approved Algorithms

Name	Algorithms/Key Lengths for use between 2011-2029 (per SP 800-57 and SP 800-131)	Algorithms/Key Lengths for use now and beyond 2030 (per SP 800-57 and SP 800-131)	References
Secure Hash Standard (SHS):	SHA-224 is Approved for all applications.	SHA-256, SHA-384, and SHA-512 are Approved for all applications.	National Institute of Standards and Technology, *Secure Hash Standard*, Federal Information Processing Standards Publication 180-3, October, 2008.
Secure Hash Algorithm (SHA)	Additionally, hash functions listed in the next column are Approved during this time.		(Note: FIPS 180-4 is expected to be released in the near future).

Table 4-4 Message Authentication – Approved Algorithms

Name	Algorithms/Key Lengths for use between 2011-2029 (per SP 800-57 and SP 800-131)	Algorithms/Key Lengths for use now and beyond 2030 (per SP 800-57 and SP 800-131)	References
CMAC	CMAC with 3-key TDES Additionally, all algorithms/key lengths listed in the next column are Approved during this time. (Note: CMAC with 2-key TDES has 80 bits of security strength. All new implementations should have 112 bits of security strength or higher.)	CMAC with AES-128, AES-192, or AES-256 (Note: CMAC with TDES is not Approved because it has <128 bits of security.)	National Institute of Standards and Technology, *Recommendation for Block Cipher Modes of Operation: The CMAC Mode for Authentication*, Special Publication 800-38B, May 2005.
CCM	All algorithms/key sizes listed in the next column are Approved during this time.	CCM with AES-128, AES-192, or AES-256	National Institute of Standards and Technology, *Recommendation for Block Cipher Modes of Operation: The CCM Mode for Authentication and Confidentiality*, Special Publication 800-38C, May

GCM/GMAC	All algorithms/key sizes listed in the next column are Approved during this time.	GCM with AES-128, AES-192, or AES-256	National Institute of Standards and Technology, *Recommendation for Block Cipher Modes of Operation: Galois/Counter Mode (GCM) and GMAC*, Special Publication 800-38D, November 2007.
HMAC	HMAC with SHA-1, SHA-224, SHA-256, SHA-384, or SHA-512 with 112≤Key Length<128 bits Additionally, all algorithms/key sizes listed in the next column are Approved during this time. (Note: 2-key TDES has 80 bits of security strength. HMAC with Key Length <112 bits is should not be used because it is being phased out by NIST. All new implementations should have 112 bits of security strength or higher.)	HMAC with SHA-1, SHA-224, SHA-256, SHA-384, or SHA-512 with Key Length≥128 bits (Note: Any HMAC with Key Length <128 bits is not Approved because it has <128 bits of security.)	National Institute of Standards and Technology, *The Keyed-Hash Message Authentication Code (HMAC)*, Federal Information Processing Standards Publication 198, March 06, 2002

Table 4-5 Key Management – Approved Algorithms

Name	Algorithms/Key Lengths for use between 2011-2029 (per SP 800-57 and SP 800-131)	Algorithms/Key Lengths for use now and beyond 2030 (per SP 800-57 and SP 800-131)	References
SP 800-108 KDFs	See rules for HMAC and CMAC; the PRFs used by the KDFs are based on these algorithms.	See rules for HMAC and CMAC; the PRFs used by the KDFs are based on these algorithms.	National Institute of Standards and Technology, *Recommendation for Key Derivation Using Pseudorandom Functions*, Special Publication 800-108, October 2009, Revised.

Table 4-6 Deterministic Random Number Generators – Approved Algorithms

Name	Algorithms/Key Lengths for use between 2011-2029 (per SP 800-57 and SP 800-131)	Algorithms/Key Lengths for use now and beyond 2030 (per SP 800-57 and SP 800-131)	References
FIPS 186-2 Appendix 3.1 RNG	FIPS 186-2 RNG will be phased out by NIST by 2015. (Note: The use of SP 800-90 RNGs is recommended since all other RNGs are being phased out by NIST.)	N/A – cannot use FIPS 186-2 RNG (Note: The use of SP 800-90 RNGs is recommended since all other RNGs are being phased out by NIST.)	National Institute of Standards and Technology, *Digital Signature Standard (DSS)*, Federal Information Processing Standards Publication 186-2, January 27, 2000 with Change Notice – Appendix 3.1.
FIPS 186-2 Appendix 3.2 RNG	FIPS 186-2 RNG will be phased out by NIST by 2015. (Note: The use of SP 800-90 RNGs is recommended since all other RNGs are being phased out by NIST.)	N/A – cannot use FIPS 186-2 RNG (Note: The use of SP 800-90 RNGs is recommended since all other RNGs are being phased out by NIST.)	National Institute of Standards and Technology, *Digital Signature Standard (DSS)*, Federal Information Processing Standards Publication 186-2, January 27, 2000 with Change Notice – Appendix 3.2. Note: Please review National Institute of Standards and Technology, *Implementation Guidance for FIPS PUB 140-1 and the Cryptographic Module Validation Program*, Sections 8.1, 8.7 and 8.9 for additional guidance.
ANSI X9.31-1998 Appendix A.2.4 RNG	ANSI X9.31 RNG will be phased out by NIST by 2015. (Note: The use of SP 800-90 RNGs is recommended since all other RNGs are being phased out by NIST.)	N/A – cannot use ANSI X9.31 RNG (Note: The use of SP 800-90 RNGs is recommended since all other RNGs are being phased out by NIST.)	American Bankers Association, *Digital Signatures Using Reversible Public Key Cryptography for the Financial Services Industry (rDSA)*, ANSI X9.31-1998 - Appendix A.2.4.
ANSI X9.62-1998 Annex A.4 RNG	ANSI X9.62-1998 RNG will be phased out by NIST by 2015. (Note: The use of SP 800-90 RNGs is recommended since all other RNGs are being phased out by NIST.)	N/A – cannot use ANSI X9.62-1998 RNG (Note: The use of SP 800-90 RNGs is recommended since all other RNGs are being phased out by NIST.)	American Bankers Association, *Public Key Cryptography for the Financial Services Industry: The Elliptic Curve Digital Signature Algorithm (ECDSA)*, ANSI X9.62-1998 – Annex A.4.

Name	Algorithms/Key Lengths for use between 2011-2029 (per SP 800-57 and SP 800-131)	Algorithms/Key Lengths for use now and beyond 2030 (per SP 800-57 and SP 800-131)	References
ANSI X9.31 Appendix A.2.4 RNG using TDES and AES RNG	ANSI X9.31 RNG will be phased out by NIST by 2015. (Note: The use of SP 800-90 RNGs is recommended since all other RNGs are being phased out by NIST.)	N/A – cannot use ANSI X9.31 RNG (Note: The use of SP 800-90 RNGs is recommended since all other RNGs are being phased out by NIST.)	National Institute of Standards and Technology, *NIST-Recommended Random Number Generator Based on ANSI X9.31 Appendix A.2.4 Using the 3-Key Triple DES and AES Algorithms*, January 31, 2005.
SP 800-90 RNG	CTR DRBG with 3-key TDES is Approved. Additionally, all algorithms/key sizes listed in the next column are Approved during this time.	HASH DRBG with SHA-1, SHA-224, SHA-256, SHA-384, or SHA-512 HMAC DRBG with SHA-1, SHA-224, SHA-256, SHA-384, or SHA-512 CTR DRBG with AES-128, AES-192, or AES-256 DUAL EC DRBG with P-256, P-384, or P-521 (Note: CTR DRBG with 3-key TDES is not Approved because it has <128 bits of security.)	National Institute of Standards and Technology, *Recommendation for Random Number Generation Using Deterministic Random Bit Generators (Revised)*, Special Publication 800-90, March 2007.

Table 4-7 Non-Deterministic Random Number Generators – Algorithms

Name	Algorithms/Key Lengths for use between 2011-2029 (per SP 800-57 and SP 800-131)	Algorithms/Key Lengths for use now and beyond 2030 (per SP 800-57 and SP 800-131)	References
Non-deterministic Random Number Generators	N/A – Currently none	N/A – Currently none	There are no FIPS Approved non-deterministic random number generators. Non-Approved RNGs may be used to seed Approved RNGs.

(Note: The requirements for Non-deterministic and Non-Approved RNGs are still an open topic. CMVP guidance may change in 2015.)

Table 4-8 Symmetric Key Establishment Techniques – Approved Algortihms

Name	Algorithms/Key Lengths for use between 2011-2029 (per SP 800-57 and SP 800-131)	Algorithms/Key Lengths for use now and beyond 2030 (per SP 800-57 and SP 800-131)	References
FIPS 140-2 IG D.2	3-key TDES Key Wrap *is* allowed. AES Key Wrap is Draft. (Note: 2-key TDES Key Wrap should not be used because it is being phased out by NIST. All new implementations should have 112 bits of security strength or higher.)	AES Key Wrap with 128-bit keys or higher *is* allowed. (Note: 3-key TDES Key Wrap is not allowed because it has <128 bits of security.)	The symmetric key establishment techniques are listed in *FIPS 140-2 Implementation Guidance* Section D.2.

Table 4-9 Asymmetric Key Establishment Techniques – Approved Algortihms

Name	Algorithms/Key Lengths for use between 2011-2029 (per SP 800-57 and SP 800-131)	Algorithms/Key Lengths for use now and beyond 2030 (per SP 800-57 and SP 800-131)	References
SP 800-56A	Key Establishment with Parameter Sets FB, FC, and EB are Approved. Key Establishment using Diffie-Hellman is approved. Additionally, all algorithms/key sizes listed in the next column	Key Establishment with Parameter Sets EC, ED, and EE are Approved.	National Institute of Standards and Technology, *Recommendation for Pair-Wise Key Establishment Schemes Using Discrete Logarithm Cryptography (Revision1)*, Special Publication 800-56A, March 2007.

Name	Algorithms/Key Lengths for use between 2011-2029 (per SP 800-57 and SP 800-131)	Algorithms/Key Lengths for use now and beyond 2030 (per SP 800-57 and SP 800-131)	References
	may be approved during this time. (Note: Parameter Sets FA and EA should not be used because they are being phased out by NIST. All new implementations should have 112 bits of security strength or higher.)	(Note: Parameter Sets FB, FC, and EB are not Approved because they have <128 bits of security.)	
SP 800-56B	Key Establishment using RSA-2048 for key transport/key agreement is Approved. (Note: RSA-1024 should not be used because it is being phased out by NIST. All new implementations should have 112 bits of security strength or higher.)	N/A – Cannot use RSA-2048 (Note: Use with RSA-2048 is not Approved because it has <128 bits of security.)	National Institute of Standards and Technology, *Recommendation for Pair-Wise Key Establishment Schemes Using Integer Factorization Cryptography*, Special Publication 800-56B, August 2009
FIPS 140-2 IG D.2	SP 800-56A primitives (using Parameter Sets FB, FC, and EB) with non-SP 800-56A KDFs in IG D.2 are allowed. Additionally, all algorithms/key sizes listed in the next column are allowed during this time. *Important:* These algorithms are only "allowed" in FIPS mode at this time. It is unclear if they will become Approved.	SP 800-56A primitives (using Parameter Sets EC, ED, and EE) with non-SP 800-56A KDFs in IG D.2 are allowed. *Important:* These algorithms are only "allowed" in FIPS mode at this time. It is unclear if	Additional asymmetric key establishment schemes are allowed in a FIPS Approved mode of operation. These schemes are listed with appropriate restrictions in *FIPS 140-2 Implementation Guidance* Section D.2.

Name	Algorithms/Key Lengths for use between 2011-2029 (per SP 800-57 and SP 800-131)	Algorithms/Key Lengths for use now and beyond 2030 (per SP 800-57 and SP 800-131)	References
	See IG D.2 for details. (Note: Parameter Sets FA and EA should not be used because they are being phased out by NIST. All new implementations should have 112 bits of security strength or higher.)	they will become Approved. See IG D.2 for details. (Note: Parameter Sets FB, FC, and EB are not allowed because they have <128 bits of security.)	

Table 4-10 Comparable Key Strengths

Bits of Security	Symmetric Key Algorithms	FCC (e.g., DSA, D-H)	IFC (e.g., RSA)	ECC (e.g., ECDSA)
80	2TDEA	L = 1024 N = 160	k = 1024	f = 160-223
112	3TDEA	L = 2048 N = 224	k = 2048	f = 224-255
128	AES-128	L = 3072 N = 256	k = 3072	f = 256-383
192	AES-192	L = 7680 N = 384	k = 7680	f = 384-511
256	AES-256	L = 15360 N = 512	k = 15360	f ≥ 512

Table 4-11 Crypto Lifetimes[33]

Algorithm Security Lifetimes	Symmetric Key Algorithms (Encryption and MAC)	FCC (e.g., DSA, D-H)	IFC (e.g., RSA)	ECC (e.g., ECDSA)
Through December 31, 2013 (minimum of 80 bits of strength)	2TDEA[a] 3TDEA AES-128 AES-192 AES-256	$\|p\| = 1024; \|q\| = 160$[b] $\|p\| \geq 2048; \|q\| \geq 224$[c]	$1024 \leq \|n\| < 2048$[d] $\|n\| \geq 2048$[e]	$160 \leq \|n\| < 224$[b] $\|n\| \geq 224$[f]
Through December 31, 2030 (minimum of 112 bits of strength)	3TDEA AES-128 AES-192 AES-256	Min: L = 2048 N = 228	Min: k = 2048	Min: f = 224
Beyond 2030 (minimum of 128 bits of strength)	AES-128 AES-192 AES-256	Min: L = 3072 N = 256	Min: k = 3072	Min: f = 256

a Encryption: acceptable through 2010; restricted use from 2011-2015. Decryption: acceptable through 2010; legacy use after 2010.

b Digital signature generation and key agreement: acceptable through 2010; deprecated from 2011 through 2013. Digital signature verification: acceptable through 2010; legacy use after 2010.

c Digital signature generation and verification: acceptable. Key agreement: $\|p\|$=2048, and $\|q\|$=224 acceptable.

d Digital signature generation: acceptable through 2010; deprecated from 2011 through 2013. Digital signature verification: acceptable through 2010; legacy use after 2010. Key agreement and key transport: $\|n\|$=1024 acceptable through 2010, and deprecated from 2011 through 2013.

e Digital signature generation and verification: acceptable. Key agreement and key transport: $\|n\|$=2048 acceptable.

f Digital signature generation and verification, and key agreement: acceptable.

---

[33] See SP 800-131 for details.

Table 4-12 Hash Function Security Strengths

Bits of Security	Digital Signatures and Hash-Only Applications	HMAC	Key Derivation Functions	Random Number Generation
80	SHA-1 SHA-224 SHA-256 SHA-384 SHA-512	SHA-1 SHA-224 SHA-256 SHA-384 SHA-512	SHA-1 SHA-224 SHA-256 SHA-384 SHA-512	SHA-1 SHA-224 SHA-256 SHA-384 SHA-512
112	SHA-224 SHA-256 SHA-384 SHA-512	SHA-1 SHA-224 SHA-256 SHA-384 SHA-512	SHA-1 SHA-224 SHA-256 SHA-384 SHA-512	SHA-1 SHA-224 SHA-256 SHA-384 SHA-512
128	SHA-256 SHA-384 SHA-512	SHA-1 SHA-224 SHA-256 SHA-384 SHA-512	SHA-1 SHA-224 SHA-256 SHA-384 SHA-512	SHA-1 SHA-224 SHA-256 SHA-384 SHA-512
192	SHA-384 SHA-512	SHA-224 SHA-256 SHA-384 SHA-512	SHA-224 SHA-256 SHA-384 SHA-512	SHA-224 SHA-256 SHA-384 SHA-512
256	SHA-512	SHA-256 SHA-384 SHA-512	SHA-256 SHA-384 SHA-512	SHA-256 SHA-384 SHA-512

4.3.3 KMS Requirements Matrix

4.3.3.1 Key Attribute Definitions

- **Key material and crypto operation protection:** A cryptography module's ability to protect its operational state from tampering and/or provide evidence of tampering. The module should also be able to keep its internal state private from general access. In the case of a Hardware Security Module (HSM), such protections are provided through physical hardware controls. In the case of software, such protection are limited and logical in nature, and may make use of some underlying hardware and operating system platform controls that offer memory protections, privileged execution states, tamper-detections, etc.

- **Key material uniqueness:** The KMS ensures that there is an adequate diversity of key material across the various devices and components participating in a system. For example, this is in order to protect against a compromise of one device such as a smart meter causing to a collapse of security in an entire system if all the keys are the same.

- **Key material generation:** The generation of key materials is secure and inline with established and known good methods, such as those listed in the NIST FIPS-140-2 standards.

- **Local autonomy:** All authentication processes between devices, or between users and devices will be able to operate even if a centralized service over a network is not available at any given time. For example, this is to ensure that if a network connection in a substation becomes unavailable, but a critical operation needs to be accomplished by local personnel, they would not in any way be inhibited from doing so.

- **Revocation management:** The ability to revoke credentials in a system in an ordered manner that ensures that all affected devices and users are notified and can take appropriate actions and adjustments to their configurations. Examples can include handling revoked PKI certificates and ensuring that entities with revoked certificates cannot be authenticated to protected services and functions.

- **Key material provisioning:** The processes and methods used to securely enter key material initially into components and devices of a system, as well as changing key materials during their operation.

- **Key material destruction:** The secure disposal of all key material after its intended use and lifetime, for example, the zeriozation / erasure of CSPs. Making key material unavailable is an acceptable alternative for systems where destruction is not possible.

- **Credential span of control:** The number of organizations, domains, systems or entities controlled or controllable through the use of the key material associated with the credential. This does not explicitly address keys used for purposes other than control nor include asymmetric keys that are indirectly used for control, such as those associated with root or intermediate certification authorities.

4.3.3.2 General Definitions

- **Hardware Security Module (HSM):** A module that provides tamper evidence/proofing, as well as the protection of all critical security parameters (CSPs) and cryptographic processes from the systems they operate in such that they can never be accessed in plaintext outside of the module.

- **Root of security:** A credential/secret or aggregation point of credentials such that there is a catastrophic loss of trust if compromised. Alternatively, root(s) of hierarchical trust credentials.

4.3.3.3 KMS Requirements

Table 4-13 KMS Requirements

Attribute	Low	Moderate	High	Requirements	Reference
Key material and cryptographic operations protection		X	X	Software protection of cryptographic materials used in individual devices (e.g. control system devices)	FIPS 140-2 Level 1
			X	Hardware protection (such as HSM) for Critical Security Parameters (CSPs) for Roots of security. It is recommended where possible to use FIPS-140-2 Level 2 or above for Physical Security.	FIPS 140-2 Levels 2 through 4
				Note: • *Symmetric and Asymmetric Keys used for authorization shall be protected from generation until the end of the cryptoperiod.* • *The integrity of all keys used for authorization must be protected. The confidentiality of Private and Symmetric keys must be protected.*	
Key material uniqueness, (e.g., key derivation secrets, managing secrets, pre-shared secrets)		X	X	Key diversity is required for High-assurance devices (unique keys per device (asymmetric) or device pairs (symmetric). This is to ensure that a single compromise of a device cannot lead to a complete collapse of security of the entire system.	NIST SP 800-57, Section 5.2
		X	X	All root key material shall be unique (with the exception of derived materials).	
Key material generation	X	X	X	Use Approved methods.	FIPS 140-2, Section 4.7.2 Annex C: Approved Random Number Generators for FIPS PUB 140-2
	X	X	X	NIST-approved RNGs need to be used.	FIPS 140-2, Section 4.7.2

Attribute	Low	Moderate	High	Requirements	Reference
				Note: There is some concern that there needs to be non-NIST approved RNG to address the lack of entropy available to some SG devices. FIPS allows the use of non-deterministic RNGs to produce entropy. Pre-loading entropy is also acceptable.}	Annex C: Approved Random Number Generators for FIPS PUB 140-2
Local autonomy (Availability Exclusively)		X	X	Must always be locally autonomous. That is no authentication process must depend on a centralized service such that if it were to become unavailable local access would not be possible.	
Revocation management	X	X	X	A credential revocation process must be established whereby all parties relying on a revoked key are informed of the revocation with complete identification of the keying material, and information that allows a proper response to the revocation.	NIST SP 800-57, Section 8.3.5
			X	Near real time/real time revocation (for example: a push based mechanism)	

Key material provisioning			X	Key distribution shall be performed in accordance with sp 800-57 (ref section 8.1.5.2.2)
				• Keys distributed manually (i.e., by other than an electronic key transport protocol) shall be protected throughout the distribution process.
				• During manual distribution, secret or private keys shall either be encrypted or be distributed using appropriate physical security procedures.
				   ○ The distribution shall be from an authorized source,
				   ○ Any entity distribution plaintext keys is trusted by both the entity that generates the keys and the entity(ies) that receives the keys,
					NIST SP 800-57, Section 8.1.5.2.2
					
					FIPS 140-2, Sections 4.7.3 and 4.7.4

Attribute	Low	Moderate	High	Requirements	Reference
				○ The keys are protected in accordance with Section 6 [800-57], and ○ The keys are received by the authorized recipient.	
	X	X	X	Keys entered over a network interface must be encrypted (not for trusted roots). *Note: This is defined for operational provisioning of a system. That is manufacture time key material is provisioned that is a bootstrap for user/owner based provisioning.*	FIPS 140-2, Section 4.7.4
			X	The manual entry of plaintext keys or key components must be performed over a trusted interface. (e.g. a dedicated, physical point to point connection to an HSM) for some higher assurance modules it will also require split or encrypted key entry.	FIPS 140-2, Section 4.7.4
Key material Destruction		X	X	All copies of the private or symmetric key shall be destroyed as soon as no longer required (e.g., for archival or reconstruction activity).	SP 800-57, Section 8.3.4
		X	X	Any media on which unencrypted keying material requiring confidentiality protection is stored shall be erased in a manner that removed all traces of the keying material so that it cannot be recovered by either physical or electronic means	SP 800-57, Section 8.3.4 FIPS 140-2, Section 4.7.6
				Note: If key destruction needs to be assured, then an HSM must be used. Zeroization applies to an operational environment and does not apply to keys that may be archived.	SP 800-57, Section 8.3.4
Key and crypto lifecycles (supersession / revocation)	X	X	X	NIST recommended cryptoperiods shall be used (SP 800-57, table 1 provides a summary) *Note: Mechanism used to replace a key must have at least the same crypto strength as the key it is replacing.*	SP 800-57, Table 1
				Note: Cryptoperiod. The requirement will be to follow SP 800-57 Key management requirements. Supersession:	

Attribute	Low	Moderate	High	Requirements	Reference
				process of creating the next key and moving to that key and getting rid of old key.	
Credential span of control		X	X	The span of control for asymmetric keys shall in general be limited to a domain or a set of contiguous domains under the control of a single legal entity such as a systems operator. Exceptions to this requirement MAY include: Root and Intermediate CAs servicing multi-system consortia where a common identity or credentialing system is required. Note: For symmetric keys, the requirement for a single pair of systems is due to the underlying requirement that the compromise of one entity should not give you control over other entities (that you didn't already have). For asymmetric keys, the underlying requirement is to be able to have a finite space in which the revocations need to be distributed.	
		X	X	A symmetric key shall not be used for control of more than a single entity.	

4.4 REFERENCES & SOURCES

1. NISTIR 7628, Draft 2: http://collaborate.nist.gov/twiki-sggrid/bin/view/SmartGrid/NISTIR7628Feb2010
2. Bottom Up Cyber Security Analysis of Smart Grid, latest version from http://collaborate.nist.gov/twiki-sggrid/bin/view/SmartGrid/CSCTGBottomUp
3. High-level requirements collection: http://collaborate.nist.gov/twiki-sggrid/bin/view/SmartGrid/CSCTGHighLevelRequirements
4. NIST Smart Grid Architecture materials: http://collaborate.nist.gov/twiki-sggrid/bin/view/SmartGrid/CsCTGArchi
5. FIPS 140-2, *Security Requirements for Cryptographic Modules*
6. FIPS 180-3, *Secure Hash Standard (SHS)*
7. FIPS 186-3, *Digital Signature Standard (DSS)*
8. FIPS 197, *Advanced Encryption Standard (AES)*
9. FIPS 198-1, *The Keyed-Hash Message Authentication Code (HMAC)*
10. NIST SP 800-21, *Guideline for Implementing Cryptography in the Federal Government*
11. NIST SP 800-22, *A Statistical Test Suite for Random and Pseudorandom Number Generators for Cryptographic Applications*
12. NIST SP 800-32, *Introduction to Public Key Technology and the Federal PKI Infrastructure*
13. NIST SP 800-56A, *Recommendation for Pair-Wise Key Establishment Schemes Using Discrete Logarithm Cryptography*
14. NIST SP 800-56B, *Recommendation for Pair-Wise Key Establishment Schemes Using Integer Factorization Cryptography*
15. NIST SP 800-57, *Recommendation for Key Management*
16. NIST SP 800-81, *Secure Domain Name System (DNS) Deployment Guide*
17. NIST SP 800-89, *Recommendation for Obtaining Assurances for Digital Signature Applications*
18. NIST SP 800-90, *Recommendation for Random Number Generation Using Deterministic Random Bit Generators*
19. NIST SP 800-102, *Recommendation for Digital Signature Timeliness*
20. SP 800-131, *DRAFT Recommendation for the Transitioning of Cryptographic Algorithms and Key Sizes*
21. American National Standard Institute, "Meter and End Device Tables communications over any network", ANSI C12.22-2008, 2008.
22. B. Aboba, et al., "Extensible Authentication Protocol (EAP)", RFC 3748, June 2004.

23. "Extensible Authentication Protocol (EAP) Key Management Framework", RFC 5247, August 2008.

24. J. Salowey, et al., "Specification for the Derivation of Root Key from an Extended Master Session Key (EMSK)", RFC 5295, August 2008.

25. National Security Agency, Suite B Cryptography, http://www.nsa.gov/ia/programs/suiteb_cryptography/index.shtml

APPENDIX A
CROSSWALK OF CYBER SECURITY DOCUMENTS

Table A-1 Crosswalk of Cyber Security Requirements and Documents

Dark Gray = Unique Technical Requirement   Light Gray = Common Technical Requirement
White = Common Governance, Risk and Compliance (GRC)

Smart Grid Cyber Security Requirement		NIST SP 800-53 Revision 3		DHS Catalog of Control Systems Security: Recommendations for Standards Developers		NERC CIPS (1-9) May 2009
\multicolumn{7}{c}{Access Control (SG.AC)}						
SG.AC-1	Access Control Policy and Procedures	AC-1	Access Control Policy and Procedures	2.15.1	Access Control Policies and Procedures	CIP 003-2 (R1, R1.1, R1.3, R5, R5.3)
SG.AC-2	Remote Access Policy and Procedures	AC-17	Remote Access	2.15.23	Remote Access Policy and Procedures	CIP005-2 (R1, R1.1, R1.2, R2, R2.3, R2.4)
SG.AC-3	Account Management	AC-2	Account Management	2.15.3	Account Management	CIP 003-2 (R5, R5.1, R5.2, R5.3) CIP 004-2 (R4, R4.1, R4.2) CIP 005-2 (R2.5) CIP 007-2 (R5, R5.1, R5.2)
SG.AC-4	Access Enforcement	AC-3	Access Enforcement	2.15.7	Access Enforcement	CIP 004-2 (R4) CIP 005-2 (R2, R2.1-R2.4)
SG.AC-5	Information Flow Enforcement	AC-4	Information Flow Enforcement	2.15.15	Information Flow Enforcement	
SG.AC-6	Separation of Duties	AC-5	Separation of Duties	2.15.8	Separation of Duties	
SG.AC-7	Least Privilege	AC-6	Least Privilege	2.15.9	Least Privilege	CIP 007-2 (R5.1)
SG.AC-8	Unsuccessful Login Attempts	AC-7	Unsuccessful Login Attempts	2.15.20	Unsuccessful Logon Notification	

Dark Gray = Unique Technical Requirement   Light Gray = Common Technical Requirement
White = Common Governance, Risk and Compliance (GRC)

Smart Grid Cyber Security Requirement		NIST SP 800-53 Revision 3		DHS Catalog of Control Systems Security: Recommendations for Standards Developers		NERC CIPS (1-9) May 2009
SG.AC-9	Smart Grid Information System Use Notification	AC-8	System Use Notification	2.15.17	System Use Notification	CIP 005-2 (R2.6)
SG.AC-10	Previous Logon Notification	AC-9	Previous Logon (Access) Notification	2.15.19	Previous Logon Notification	
SG.AC-11	Concurrent Session Control	AC-10	Concurrent Session Control	2.15.18	Concurrent Session Control	
SG.AC-12	Session Lock	AC-11	Session Lock	2.15.21	Session Lock	
SG.AC-13	Remote Session Termination			2.15.22	Remote Session Termination	
SG.AC-14	Permitted Actions without Identification or Authentication	AC-14	Permitted Actions without Identification or Authentication	2.15.11	Permitted Actions without Identification and Authentication	
SG.AC-15	Remote Access	AC-17	Remote Access	2.15.24	Remote Access	CIP 005-2 (R2, R3, R3.1, R3.2)
SG.AC-16	Wireless Access Restrictions			2.15.26	Wireless Access Restrictions	
SG.AC-17	Access Control for Portable and Mobile Devices	AC-19	Access Control for Mobile Devices	2.15.25	Access Control for Portable and Mobile Devices	CIP 005-2 (R2.4, R5, R5.1)
SG.AC-18	Use of External Information Control Systems	SC-7	Boundary Protection	2.15.29	Use of External Information Control Systems	
SG.AC-19	Control System Access Restrictions			2.15.28	External Access Protections	
SG.AC-20	Publicly Accessible Content					
SG.AC-21	Passwords			2.15.16	Passwords	CIP 007-2 (R5.3)

Awareness and Training (SG.AT)

Dark Gray = Unique Technical Requirement Light Gray = Common Technical Requirement
White = Common Governance, Risk and Compliance (GRC)

Smart Grid Cyber Security Requirement			NIST SP 800-53 Revision 3		DHS Catalog of Control Systems Security: Recommendations for Standards Developers	NERC CIPS (1-9) May 2009
SG.AT-1	Awareness and Training Policy and Procedures	AT-1	Security Awareness and Training Policy and Procedures	2.11.1	Security Awareness Training Policy and Procedures	CIP 004-2 (R1, R2)
SG.AT-2	Security Awareness	AT-2	Security Awareness	2.11.2	Security Awareness	CIP 004-2 (R1)
SG.AT-3	Security Training	AT-3	Security Training	2.11.3	Security Training	CIP 004-2 (R2)
SG.AT-4	Security Awareness and Training Records	AT-4	Security Training Records	2.11.4	Security Training Records	CIP 004-2 (R2.3)
SG.AT-5	Contact with Security Groups and Associations	AT-5	Contact with Security Groups and Associations	2.11.5	Contact with Security Groups and Associations	
SG.AT-6	Security Responsibility Training			2.11.6	Security Responsibility Training	
SG.AT-7	Planning Process Training			2.7.5	Planning Process Training	CIP 004-2 (R2)
Audit and Accountability (SG.AU)						
SG.AU-1	Audit and Accountability	AU-1	Audit and Accountability Policy and Procedures	2.16.1	Audit and Accountability Process and Procedures	CIP 003-2 (R1, R1.1, R1.3)
SG.AU-2	Auditable Events	AU-2	Auditable Events	2.16.2	Auditable Events	CIP 005-2 (R1, R1.1, R1.3) CIP 007-2 (R5.1.2, R5.2.3, R6.1, R6.3)
		AU-13	Monitoring for Information Disclosure			
SG.AU-3	Content of Audit Records	AU-3	Content of Audit Records	2.16.3	Content of Audit Records	CIP 007-3 (R5.1.2)
SG.AU-4	Audit Storage Capacity	AU-4	Audit Storage Capacity	2.16.4	Audit Storage	

Dark Gray = Unique Technical Requirement Light Gray = Common Technical Requirement
White = Common Governance, Risk and Compliance (GRC)

Smart Grid Cyber Security Requirement		NIST SP 800-53 Revision 3		DHS Catalog of Control Systems Security: Recommendations for Standards Developers		NERC CIPS (1-9) May 2009
SG.AU-5	Response to Audit Processing Failures	AU-5	Response to Audit Processing Failures	2.16.5	Response to Audit Processing Failures	
SG.AU-6	Audit Monitoring, Analysis, and Reporting	AU-6	Audit Monitoring, Analysis, and Reporting	2.16.6	Audit Monitoring, Process, and Reporting	CIP 007-2 (R5.1.2) CIP 007-2 (R6.5)
SG.AU-7	Audit Reduction and Report Generation	AU-7	Audit Reduction and Report Generation	2.16.7	Audit Reduction and Report Generation	
SG.AU-8	Time Stamps	AU-8	Time Stamps	2.16.8	Time Stamps	
SG.AU-9	Protection of Audit Information	AU-9	Protection of Audit Information	2.16.9	Protection of Audit Information	CIP 003-2 (R4)
SG.AU-10	Audit Record Retention	AU-11	Audit Record Retention	2.16.10	Audit Record Retention	CIP 005-2 (R5.3) CIP 007-2 (R5.1.2, R6.4) CIP 008-2 (R2)
SG.AU-11	Conduct and Frequency of Audits	AU-1	Audit and Accountability Policy and Procedures	2.16.11	Conduct and Frequency of Audits	
SG.AU-12	Auditor Qualification			2.16.12	Auditor Qualification	
SG.AU-13	Audit Tools	AU-7	Audit Reduction and Report Generation	2.16.13	Audit Tools	
SG.AU-14	Security Policy Compliance	CA-1	Security Assessment and Authorization Policies and Procedures	2.16.14	Security Policy Compliance	
SG.AU-15	Audit Generation	AU-12	Audit Generation	2.16.15	Audit Generation	
SG.AU-16	Non-Repudiation	AU-10	Non-Repudiation	2.16.16	Non-Repudiation	

Security Assessment and Authorization (SG.CA)

Dark Gray = Unique Technical Requirement   Light Gray = Common Technical Requirement
White = Common Governance, Risk and Compliance (GRC)

Smart Grid Cyber Security Requirement		NIST SP 800-53 Revision 3		DHS Catalog of Control Systems Security: Recommendations for Standards Developers		NERC CIPS (1-9) May 2009
SG.CA-1	Security Assessment and Authorization Policy and Procedures	CA-1	Security Assessment and Authorization Policies and Procedures	2.18.3	Certification, Accreditation, and Security Assessment Policies and Procedures	
				2.17.1	Monitoring and Reviewing Control System Security management Policy and Procedures	
SG.CA-2	Security Assessments	CA-2	Security Assessments	2.17.3	Monitoring of Security Policy	
SG.CA-3	Continuous Improvement			2.17.2	Continuous Improvement	
				2.17.4	Best Practices	
SG.CA-4	Information System Connections	CA-3	Information System Connection	2.18.5	Control System Connections	CIP 005-2 (R2)
SG.CA-5	Security Authorization to Operate	CA-6	Security Authorization	2.17.5	Security Accreditation	
		PM-10	Security Authorization Process			
SG.CA-6	Continuous Monitoring	CA-7	Continuous Monitoring	2.18.7	Continuous Monitoring	
Configuration Management (SG.CM)						
SG.CM-1	Configuration Management Policy and Procedures	CM-1	Configuration Management Policy and Procedures	2.6.1	Configuration Management Policy and Procedures	CIP 003-2 (R6)
SG.CM-2	Baseline Configuration	CM-2	Baseline Configuration	2.6.2	Baseline Configuration	CIP 007-2 (R9)
SG.CM-3	Configuration Change Control	CM-3	Configuration Change Control	2.6.3	Configuration Change Control	CIP 003-2 (R6)

NISTIR 7628 Guidelines for Smart Grid Cyber Security v1.0 – Aug 2010

Dark Gray = Unique Technical Requirement Light Gray = Common Technical Requirement
White = Common Governance, Risk and Compliance (GRC)

Smart Grid Cyber Security Requirement		NIST SP 800-53 Revision 3		DHS Catalog of Control Systems Security: Recommendations for Standards Developers		NERC CIPS (1-9) May 2009
SG.CM-4	Monitoring Configuration Changes	SA-10	Developer Configuration Management		Monitoring Configuration Changes	CIP 003-2 (R6)
		CM-4	Security Impact Analysis	2.6.4		
		SA-10	Developer Configuration Management			
SG.CM-5	Access Restrictions for Configuration Change	CM-5	Access Restrictions for Change	2.6.5	Access Restrictions for Configuration Change	CIP 003-2 (R6)
SG.CM-6	Configuration Settings	CM-6	Configuration Settings	2.6.6	Configuration Settings	CIP 003-2 (R6) CIP 005 (R2.2)
SG.CM-7	Configuration for Least Functionality	CM-7	Least Functionality	2.6.7	Configuration for Least Functionality	
SG.CM-8	Component Inventory	CM-8	Information System Component Inventory	2.6.8	Configuration Assets	
SG.CM-9	Addition, Removal, and Disposal of Equipment	MP-6	Media Sanitization	2.6.9	Addition, Removal, and Disposition of Equipment	CIP 003-2 (R6)
SG.CM-10	Factory Default Settings Management			2.6.10	Factory Default Authentication Management	CIP 005-2 (R4.4)
SG.CM-11	Configuration Management Plan	CM-9	Configuration Management Plan			
Continuity of Operations (SG.CP)						
SG.CP-1	Continuity of Operations Policy and Procedures	CP-1	Contingency Planning Policy and Procedures			

Dark Gray = Unique Technical Requirement Light Gray = Common Technical Requirement
White = Common Governance, Risk and Compliance (GRC)

	Smart Grid Cyber Security Requirement		NIST SP 800-53 Revision 3		DHS Catalog of Control Systems Security: Recommendations for Standards Developers	NERC CIPS (1-9) May 2009
SG.CP-2	Continuity of Operations Plan	CP-1	Contingency Planning Policy and Procedures	2.12.2	Continuity of Operations Plan	CIP 008-2 (R1) CIP 009-2 (R1)
SG.CP-3	Continuity of Operations Roles and Responsibilities	CP-2	Contingency Plan	2.12.3	Continuity of Operations Roles and Responsibilities	CIP 009-2 (R1.1, R1.2)
SG.CP-4	Continuity of Operations Training					
SG.CP-5	Continuity of Operations Plan Testing	CP-4	Contingency Plan Testing and Exercises	2.12.5	Continuity of Operations Plan Testing	CIP 008-2 (R1.6) CIP 009-2 (R2, R5)
SG.CP-6	Continuity of Operations Plan Update			2.12.6	Continuity of Operations Plan Update	CIP 009-2 (R4, R5)
SG.CP-7	Alternate Storage Sites	CP-6	Alternate Storage Sites	2.12.13	Alternative Storage Sites	
SG.CP-8	Alternate Telecommunication Services	CP-8	Telecommunications Services	2.12.14	Alternate Command/Control Methods	
SG.CP-9	Alternate Control Center	CP-7	Alternate Processing Site	2.12.15	Alternate Control Center	
		CP-8	Telecommunications Services			
SG.CP-10	Smart Grid Information System Recovery and Reconstitution	CP-10	Information System Recovery and Reconstitution	2.12.17	Control System Recovery and Reconstitution	CIP 009-2 (R4)
SG.CP-11	Fail-Safe Response			2.12.18	Fail-Safe Response	
	Identification and Authentication (SG.IA)					
SG.IA-1	Identification and Authentication Policy	IA-1	Identification and Authentication Policy and	2.15.2	Identification and Authentication	

NISTIR 7628 Guidelines for Smart Grid Cyber Security v1.0 – Aug 2010

Dark Gray = Unique Technical Requirement Light Gray = Common Technical Requirement
White = Common Governance, Risk and Compliance (GRC)

Smart Grid Cyber Security Requirement		NIST SP 800-53 Revision 3		DHS Catalog of Control Systems Security: Recommendations for Standards Developers		NERC CIPS (1-9) May 2009
	and Procedures		Procedures		Procedures and Policy	
SG.IA-2	Identifier Management	IA-4	Identifier Management	2.15.4	Identifier Management	
SG.IA-3	Authenticator Management	IA-5	Authenticator Management	2.15.5	Authenticator Management	CIP 007-2 (R5, R5.1, R5.2, R5.3)
SG.IA-4	User Identification and Authentication	IA-2	User Identification and Authentication	2.15.10	User Identification and Authentication	CIP 003-2 (R1, R1.1, R1.3)
SG.IA-5	Device Identification and Authentication	IA-3	Device Identification and Authentication	2.15.12	Device Authentication and Identification	
SG.IA-6	Authenticator Feedback	IA-6	Authenticator Feedback	2.15.13	Authenticator Feedback	
Information and Document Management (SG.ID)						
SG.ID-1	Information and Document Management Policy and Procedures			2.9.1	Information and Document Management Policy and Procedures	
SG.ID-2	Information and Document Retention			2.9.2	Information and Document Retention	CIP 006-2 (R7)
SG.ID-3	Information Handling	MP-1	Media Protection Policy and Procedures	2.9.3	Information Handling	CIP 003-2 (R4.1)
SG.ID-4	Information Exchange			2.9.5	Information Exchange	
SG.ID-5	Automated Labeling			2.9.11	Automated Labeling	
Incident Response (SG.IR)						
SG.IR-1	Incident Response Policy and Procedures	IR-1	Incident Response Policy and Procedures	2.12.1	Incident Response Policy and Procedures	

Dark Gray = Unique Technical Requirement   Light Gray = Common Technical Requirement
White = Common Governance, Risk and Compliance (GRC)

Smart Grid Cyber Security Requirement		NIST SP 800-53 Revision 3		DHS Catalog of Control Systems Security: Recommendations for Standards Developers		NERC CIPS (1-9) May 2009
SG.IR-2	Incident Response Roles and Responsibilities	IR-1	Incident Response Policy and Procedures	2.7.4	Roles and Responsibilities	CIP 008-2 (Rr1.2) CIP 009-2 (R1.2)
SG.IR-3	Incident Response Training	IR-2	Incident Response Training	2.12.4	Incident Response Training	
SG.IR-4	Incident Response Testing and Exercises	IR-3	Incident Response Testing and Exercises			
SG.IR-5	Incident Handling	IR-4	Incident Handling	2.12.7	Incident Handling	
SG.IR-6	Incident Monitoring	IR-5	Incident Monitoring	2.12.8	Incident Monitoring	
SG.IR-7	Incident Reporting	IR-6	Incident Reporting	2.12.9	Incident Reporting	
SG.IR-8	Incident Response Investigation and Analysis	PE-6	Monitoring Physical Access	2.12.11	Incident Response Investigation and Analysis	CIP 008-2 (R1, R1.2-R1.5)
SG.IR-9	Corrective Action			2.12.12	Corrective Action	CIP 008-2 (R1.4) CIP 009-2 (R3)
SG.IR-10	Smart Grid Information System Backup	CP-9	Information System Backup	2.12.16	Control System Backup	
SG.IR-11	Coordination of Emergency Response			2.2.4	Coordination of Threat Mitigation	CIP 008-2 (R1.3)
Smart Grid Information System Development and Maintenance (SG.MA)						
SG.MA-1	Smart Grid Information System Maintenance Policy and Procedures	MA-1	System Maintenance Policy and Procedures	2.10.1	System Maintenance Policy and Procedures	
SG.MA-2	Legacy Smart Grid Information System Updates			2.10.2	Legacy System Upgrades	

Dark Gray = Unique Technical Requirement   Light Gray = Common Technical Requirement
White = Common Governance, Risk and Compliance (GRC)

Smart Grid Cyber Security Requirement		NIST SP 800-53 Revision 3		DHS Catalog of Control Systems Security: Recommendations for Standards Developers		NERC CIPS (1-9) May 2009
SG.MA-3	Smart Grid Information System Maintenance	PL-6	Security-Related Activity Planning	2.10.5	Unplanned System Maintenance	
		MA-2	Controlled Maintenance	2.10.6	Periodic System Maintenance	
SG.MA-4	Maintenance Tools	MA-3	Maintenance Tools	2.10.7	Maintenance Tools	
SG.MA-5	Maintenance Personnel	MA-5	Maintenance Personnel	2.10.8	Maintenance Personnel	
SG.MA-6	Remote Maintenance	MA-4	Non-Local Maintenance	2.10.9	Remote Maintenance	
SG.MA-7	Timely Maintenance	MA-6	Timely Maintenance	2.10.10	Timely Maintenance	CIP 009-2 (R4)
Media Protection (SG.MP)						
SG.MP-1	Media Protection Policy and Procedures	MP-1	Media Protection Policy and Procedures	2.13.1	Media Protection and Procedures	
SG.MP-2	Media Sensitivity Level	RA-2	Security Categorization	2.13.3	Media Classification	CIP 003-2 (R4, R4.2)
				2.9.4	Information Classification	
SG.MP-3	Media Marketing	MP-3	Media Marketing	2.13.4	Media Labeling	
				2.9.10	Automated Marking	
SG.MP-4	Media Storage	MP-4	Media Storage	2.13.5	Media Storage	
SG.MP-5	Media Transport	MP-5	Media Transport	2.13.6	Media Transport	CIP 007-2 (R7, R7.1, R7.2, R7.3)
SG.MP-6	Media Sanitization and Disposal	MP-6	Media Sanitization	2.13.7	Media Sanitization and Storage	
Physical and Environmental Security (SG.PE)						
SG.PE-1	Physical and Environmental Security Policy and Procedures	PE-1	Physical and Environmental Protection Policy and Procedures	2.4.1	Physical and Environmental Security Policies and Procedures	CIP 006-2 (R1, R2)

NISTIR 7628 Guidelines for Smart Grid Cyber Security v1.0 – Aug 2010

Dark Gray = Unique Technical Requirement Light Gray = Common Technical Requirement
White = Common Governance, Risk and Compliance (GRC)

Smart Grid Cyber Security Requirement		NIST SP 800-53 Revision 3		DHS Catalog of Control Systems Security: Recommendations for Standards Developers		NERC CIPS (1-9) May 2009
SG.PE-2	Physical Access Authorizations	PE-2	Physical Access Authorizations	2.4.2	Physical Access Authorizations	CIP 004-2 (R4)
SG.PE-3	Physical Access	PE-3	Physical Access Control	2.4.3	Physical Access Control	CIP 006-2 (R2)
		PE-4	Access Control for Transmission Medium			
		PE-5	Access Control for Output Devices			
SG.PE-4	Monitoring Physical Access	PE-6	Monitoring Physical Access	2.4.4	Monitoring Physical Access	CIP 006-2 (R5)
SG.PE-5	Visitor Control	PE-7	Visitor Control	2.4.5	Visitor Control	CIP 006-2 (R1.4)
SG.PE-6	Visitor Records	PE-8	Access Records	2.4.6	Visitor Records	CIP 006-2 (R1.4, R6)
SG.PE-7	Physical Access Log Retention			2.4.7	Physical Access Log Retention	CIP 006-2 (R7)
SG.PE-8	Emergency Shutoff Protection	PE-10	Emergency Shutoff	2.4.8	Emergency Shutoff	
SG.PE-9	Emergency Power	PE-11	Emergency Power	2.4.9	Emergency Power	
SG.PE-10	Delivery and Removal	PE-16	Delivery and Removal	2.4.14	Delivery and Removal	
SG.PE-11	Alternate Work Site	PE-17	Alternate Work Site	2.4.15	Alternate Work Site	
SG.PE-12	Location of Smart Grid Information System Assets	PE-18	Location of Information System Components	2.4.18	Location of Control System Assets	
Planning (SG.PL)						
SG.PL-1	Strategic Planning Policy and Procedures	PL-1	Security Planning and Procedures	2.7.1	Strategic Planning Policy and Procedures	

A-11

NISTIR 7628 Guidelines for Smart Grid Cyber Security v1.0 – Aug 2010

Dark Gray = Unique Technical Requirement   Light Gray = Common Technical Requirement
White = Common Governance, Risk and Compliance (GRC)

Smart Grid Cyber Security Requirement		NIST SP 800-53 Revision 3		DHS Catalog of Control Systems Security: Recommendations for Standards Developers		NERC CIPS (1-9) May 2009
SG.PL-2	Smart Grid Information System Security Plan	PL-2	System Security Plan	2.7.2	Control System Security Plan	
SG.PL-3	Rules of Behavior	PL-4	Rules of Behavior	2.7.11	Rules of Behavior	
SG.PL-4	Privacy Impact Assessment	PL-5	Privacy Impact Assessment			
SG.PL-5	Security-Related Activity Planning	PL-6	Security-Related Activity Planning	2.7.12	Security-Related Activity Planning	CIP 002-2 (R1)
Security Program Management (SG.PM)						
SG.PM-1	Security Policy and Procedures	AC-1	Access Control Policy and Procedures	2.1.1	Security Policies and Procedures	CIP 003-2 (R1, R1.1, R1.3, R5, R5.3)
SG.PM-2	Security Program Plan	PM-1	Information Security Program Plan			
SG.PM-3	Senior Management Authority	PM-2	Senior Information Security Officer			
SG.PM-4	Security Architecture	PM-7	Enterprise Architecture			
SG.PM-5	Risk Management Strategy	PM-9	Risk Management Strategy			
SG.PM-6	Security Authorization to Operate Process	PM-10	Security Authorization Process			
SG.PM-7	Mission/Business Process Definition	PM-11	Mission/Business Process Definition			
SG.PM-8	Management Accountability	PM-1	Information Security Program Plan	2.2.2	Management Accountability	CIP 003-2 (R2, R3)
Personnel Security (SG.PS)						

NISTIR 7628 Guidelines for Smart Grid Cyber Security v1.0 – Aug 2010

Dark Gray = Unique Technical Requirement Light Gray = Common Technical Requirement
White = Common Governance, Risk and Compliance (GRC)

Smart Grid Cyber Security Requirement		NIST SP 800-53 Revision 3		DHS Catalog of Control Systems Security: Recommendations for Standards Developers		NERC CIPS (1-9) May 2009
SG.PS-1	Personnel Security Policy and Procedures	PS-1	Personnel Security Policy and Procedures	2.3.1	Personnel Security Policies and Procedures	CIP 004-2 (R3)
SG.PS-2	Position Categorization	PS-2	Position Categorization	2.3.2	Position Categorization	CIP 004-2 (R3)
SG.PS-3	Personnel Screening	PS-3	Personnel Screening	2.3.3	Personnel Screening	CIP 004-2 (R3)
SG.PS-4	Personnel Termination	PS-4	Personnel Termination	2.3.4	Personnel Termination	CIP 004-2 (R4.2) CIP 004-2 (R5.2.3)
SG.PS-5	Personnel Transfer	PS-5	Personnel Transfer	2.3.5	Personnel Transfer	CIP 004-2 (R4.1, R4.2)
SG.PS-6	Access Agreements	PS-6	Access Agreements	2.3.6	Access Agreements	
SG.PS-7	Contractor and Third-Party Personnel Security	PS-7	Third-Party Personnel Security	2.3.7	Third-Party Security Agreements	CIP 004-2 (R3.3)
SG.PS-8	Personnel Accountability	PS-8	Personnel Sanctions	2.3.8	Personnel Accountability	
SG.PS-9	Personnel Roles			2.3.9	Personnel Roles	
Risk Management and Assessment (SG.RA)						
SG.RA-1	Risk Assessment Policy and Procedures	RA-1	Risk Assessment Policy and Procedures	2.18.1	Risk Assessment Policy and Procedures	CIP 002-2 (R1, R1.1, R1.2, R4) CIP 003-2 (R1, R4.2)
SG.RA-2	Risk Management Plan	PM-9	Risk Management Strategy	2.18.2	Risk Management Plan	CIP 003-2 (R4, R4.1, R4.2)
SG.RA-3	Security Impact Level	RA-2	Security Categorization	2.18.8	Security Categorization	
SG.RA-4	Risk Assessment	RA-3	Risk Assessment	2.18.9	Risk Assessment	CIP 002-2 (R1.2)
SG.RA-5	Risk Assessment Update	RA-3	Risk Assessment	2.18.10	Risk Assessment Update	CIP 002-2 (R4)

A-13

Dark Gray = Unique Technical Requirement   Light Gray = Common Technical Requirement
White = Common Governance, Risk and Compliance (GRC)

Smart Grid Cyber Security Requirement		NIST SP 800-53 Revision 3		DHS Catalog of Control Systems Security: Recommendations for Standards Developers		NERC CIPS (1-9) May 2009
SG.RA-6	Vulnerability Assessment and Awareness	RA-5	Vulnerability Scanning	2.18.11	Vulnerability Assessment and Awareness	CIP 005-2 (R4, R4.2, R4.3, R4.4) CIP 007-2 (R8)
Smart Grid Information System and Services Acquisition (SG.SA)						
SG.SA-1	Smart Grid Information System and Services Acquisition Policy and Procedures	SA-1	System and Services Acquisition Policy and Procedures	2.5.1	System and Services Acquisition Policy and Procedures	
SG.SA-2	Security Policies for Contractors and Third Parties			2.2.5	Security Policies for Third Parties	
				2.2.6	Termination of Third-Party Access	
SG.SA-3	Life-Cycle Support	SA-3	Life-Cycle Support	2.5.3	Life-Cycle Support	
SG.SA-4	Acquisitions	SA-4	Acquisitions	2.5.4	Acquisitions	
SG.SA-5	Smart Grid Information System Documentation	SA-5	Information System Documentation	2.5.5	Control System Documentation	
SG.SA-6	Software License Usage Restrictions	SA-6	Software Usage Restrictions	2.5.6	Software License Usage Restrictions	
SG.SA-7	User-Installed Software	SA-7	User-Installed Software	2.5.7	User-installed Software	
SG.SA-8	Security Engineering Principles	SA-8	Security Engineering Principles	2.5.8	Security Engineering Principals	
		SA-13	Trustworthiness			
SG.SA-9	Developer Configuration Management	SA-10	Developer Configuration Management	2.5.10	Vendor Configuration Management	

Dark Gray = Unique Technical Requirement   Light Gray = Common Technical Requirement
White = Common Governance, Risk and Compliance (GRC)

Smart Grid Cyber Security Requirement		NIST SP 800-53 Revision 3		DHS Catalog of Control Systems Security: Recommendations for Standards Developers		NERC CIPS (1-9) May 2009
SG.SA-10	Developer Security Testing	SA-11	Developer Security Testing	2.5.11	Vendor Security Testing	
SG.SA-11	Supply Chain Protection	SA-12	Supply Chain Protection	2.5.12	Vendor Life-cycle Practices	
Smart Grid Information System and Communication Protection (SG.SC)						
SG.SC-1	System and Communication Protection Policy and Procedures	SC-1	System and Communication Protection Policy and Procedures	2.8.1	System and Communication Protection Policy and Procedures	CIP 003-2 (R1, R1.1, R1.3)
SG.SC-2	Communications Partitioning			2.8.2	Management Port Partitioning	
SG.SC-3	Security Function Isolation	SC-3	Security Function Isolation	2.8.3	Security Function Isolation	
SG.SC-4	Information Remnants	SC-4	Information in Shared Resources	2.8.4	Information Remnants	
SG.SC-5	Denial-of-Service Protection	SC-5	Denial-of-Service Protection	2.8.5	Denial-of-Service Protection	
SG.SC-6	Resource Priority	SC-6	Resource Priority	2.8.6	Resource Priority	
SG.SC-7	Boundary Protection	SC-7	Boundary Protection	2.8.7	Boundary Protection	CIP 005-2 (R1, R1.1, R1.2, R1.3, R1.4, R1.6, R2, R2.1-R2.4, R5, R5.1)
SG.SC-8	Communication Integrity	SC-8	Transmission Integrity	2.8.8	Communication Integrity	
SG.SC-9	Communication Confidentiality	SC-9	Transmission Confidentiality	2.8.9	Communication Confidentiality	
SG.SC-10	Trusted Path	SC-11	Trusted Path	2.8.10	Trusted Path	
SG.SC-11	Cryptographic Key Establishment and Management	SC-12	Cryptographic Key Establishment and Management	2.8.11	Cryptographic Key Establishment and Management	

NISTIR 7628 Guidelines for Smart Grid Cyber Security v1.0 – Aug 2010

Dark Gray = Unique Technical Requirement   Light Gray = Common Technical Requirement
White = Common Governance, Risk and Compliance (GRC)

Smart Grid Cyber Security Requirement		NIST SP 800-53 Revision 3		DHS Catalog of Control Systems Security: Recommendations for Standards Developers		NERC CIPS (1-9) May 2009
SG.SC-12	Use of Validated Cryptography	SC-13	Use of Cryptography	2.8.12	Use of Validated Cryptography	
SG.SC-13	Collaborative Computing	SC-15	Collaborative Computing Devices	2.8.13	Collaborative Computing	
SG.SC-14	Transmission of Security Parameters	SC-16	Transmission of Security Attributes	2.8.14	Transmission of Security Parameters	
SG.SC-15	Public Key Infrastructure Certificates	SC-17	Public Key Infrastructure Certificates	2.8.15	Public Key Infrastructure Certificates	
SG.SC-16	Mobile Code	SC-18	Mobile Code	2.8.16	Mobile Code	
SG.SC-17	Voice-Over Internet Protocol	SC-19	Voice Over Internet Protocol	2.8.17	Voice-over-Internet Protocol	
SG.SC-18	System Connections	CA-3	Information System Connections	2.8.18	System Connections	CIP 005-2 (R2, R2.2-R2.4)
SG.SC-19	Security Roles	SA-9	External Information System Services	2.8.19	Security Roles	CIP 003-2 (R5)
SG.SC-20	Message Authenticity	SC-8	Transmission Integrity	2.8.20	Message Authenticity	
SG.SC-21	Secure Name/Address Resolution Service	SC-20	Secure Name/Address Resolution Service (Authoritative Source)	2.8.22	Secure Name/Address Resolution Service (Authoritative Source)	
SG.SC-22	Fail in Known State	SC-24	Fail in Known State	2.8.24	Fail in Know State	
SG.SC-23	Thin Nodes	SC-25	Thin Nodes	2.8.25	Thin Nodes	
SG.SC-24	Honeypots	SC-26	Honeypots	2.8.26	Honeypots	
SG.SC-25	Operating System-Independent Applications	SC-27	Operating System-Independent Applications	2.8.27	Operating System-Independent Applications	
SG.SC-26	Confidentiality of Information at Rest	SC-28	Confidentiality of Information at Rest	2.8.28	Confidentiality of Information at Rest	
SG.SC-27	Heterogeneity	SC-29	Heterogeneity	2.8.29	Heterogeneity	

Dark Gray = Unique Technical Requirement Light Gray = Common Technical Requirement
White = Common Governance, Risk and Compliance (GRC)

Smart Grid Cyber Security Requirement		NIST SP 800-53 Revision 3		DHS Catalog of Control Systems Security: Recommendations for Standards Developers		NERC CIPS (1-9) May 2009
SG.SC-28	Virtualization Technique	SC-30	Virtualization Technique	2.8.30	Virtualization Techniques	
SG.SC-29	Application Partitioning			2.8.32	Application Partitioning	
SG.SC-30	Information System Partitioning	SC-32	Information Systems Partitioning			
Smart Grid Information System and Information Integrity (SG.SI)						
SG.SI-1	System and Information Integrity Policy and Procedures	SI-1	System and Information Integrity Policy and Procedures	2.14.1	System and Information Integrity Policy and Procedures	
SG.SI-2	Flaw Remediation	SI-2	Flaw Remediation	2.14.2	Flaw Remediation	CIP 007-2 (R3, R3.1, R3.2)
SG.SI-3	Malicious Code and Spam Protection	SI-3	Malicious Code Protection	2.14.3	Malicious Code Protection	CIP 007-2 (R4, R4.1, R4.2)
		SI-8	Spam Protection	2.14.8	Spam Protection	CIP 007-2 (R4)
SG.SI-4	Smart Grid Information System Monitoring Tools and Techniques	SI-4	Information System Monitoring	2.14.4	System Monitoring Tools and Techniques	CIP 007-2 (R6)
SG.SI-5	Security Alerts and Advisories	SI-5	Security Alerts, Advisories, and Directives	2.14.5	Security Alerts and Advisories	
SG.SI-6	Security Functionality Verification	SI-6	Security Functionality Verification	2.14.6	Security Functionality Verification	CIP 007-2 (R1)
SG.SI-7	Software and Information Integrity	SI-7	Software and Information Integrity	2.14.7	Software and Information Integrity	
SG.SI-8	Information Input Validation	SI-10	Information Input Validation	2.14.9	Information Input Restrictions	CIP 003-2 (R5) CIP 007-2 (R, R5.1, R5.2)

Dark Gray = Unique Technical Requirement   Light Gray = Common Technical Requirement
White = Common Governance, Risk and Compliance (GRC)

Smart Grid Cyber Security Requirement		NIST SP 800-53 Revision 3		DHS Catalog of Control Systems Security: Recommendations for Standards Developers		NERC CIPS (1-9) May 2009
				2.14.10	Information Input Accuracy, Completeness, Validity and Authenticity	
SG.SI-9	Error Handling	SI-11	Error Handling	2.14.11	Error Handling	

APPENDIX B
EXAMPLE SECURITY TECHNOLOGIES AND SERVICES TO MEET THE HIGH-LEVEL SECURITY REQUIREMENTS

Power system operations have been managing the reliability of the power grid for decades in which availability of power has been a major requirement, with the integrity of information as a secondary but increasingly critical requirement. Confidentiality of customer information has also been important in the normal revenue billing processes. Although focused on inadvertent security problems, such as equipment failures, careless employees, and natural disasters, many of the existing methods and technologies can be expanded to address deliberate cyber security attacks and security compromises resulting from the expanded use of IT and telecommunications in the electric sector.

One of the most important security solutions is to utilize and augment existing power system technologies to address new risks associated with the Smart Grid. These power system management technologies (e.g., SCADA systems, EMS, contingency analysis applications, and fault location, isolation, and restoration functions, as well as revenue protection capabilities) have been refined for years to address the increasing reliability requirements and complexity of power system operations. These technologies are designed to detect anomalous events, notify the appropriate personnel or systems, continue operating during an incident/event, take remedial actions, and log all events with accurate timestamps.

In the past, there has been minimal need for distribution management except for load shedding to avoid serious problems. In the future, with generation, storage, and load on the distribution grid, utilities will need to implement more sophisticated powerflow-based applications to manage the distribution grid. Also, AMI systems can be used to provide energy-related information and act as secondary sources of information. These powerflow-based applications and AMI systems could be designed to address security.

Finally, metering has addressed concerns about confidentiality of revenue and customer information for many years. The implementation of smart meters has increased those concerns. However, many of the same concepts for revenue protection could also be used for the Smart Grid. To summarize, expanding existing power system management capabilities to cover specific security requirements, such as power system reliability, is an important area for future analysis.

Following are existing power system capabilities and features that may address the cyber security requirements included in this report. These existing capabilities may need to be tailored or expanded to meet the security requirements.

B.1 POWER SYSTEM CONFIGURATIONS AND ENGINEERING STRATEGIES

- Networked transmission grid so the loss of a single power system element will not cause a transmission outage (n-1 contingency),

- Redundant[34] power system equipment (e.g., redundant transmission lines, redundant transformers),
- Redundant information sources (e.g., redundant sensors, voltage measurements from different substation equipment or from different substations),
- Redundant communication networks (e.g., fiber optic network and power line carrier between substations, or redundant communication "headends"),
- Redundant automation systems (e.g., redundant substation protective relays, redundant SCADA computers systems, backup systems that can be quickly switched in),
- Redundant or backup control centers (e.g., SCADA systems in physically different locations),
- Redundant power system configurations (e.g., networked grids, multiple feeds to customer site from different substations),
- Redundant logs and databases with mirrored or frequent updates,
- Multiple generators connected at different locations on the transmission grid,
- Reserve generation capacity available to handle the loss of a generator,
- Configuration setting development procedures, including remedial relay settings, and
- Post-event engineering forensic analysis.

B.2 Local Equipment Monitoring, Analysis, and Control

- Sensors on substation and feeder equipment monitor volts, VARs, current, temperature, vibrations, etc. – eyes and ears for monitoring the power system,
- Control capabilities for local control, either automatically (e.g., breaker trip) or manually (e.g., substation technician raises the voltage setting on a tap changer),
- Voltage/VAR regulation by local equipment to ensure voltages and VARs remain within prescribed limits,
- Protective relaying to respond to system events (e.g., power system fault) by tripping breakers,
- Reclosers which reconnect after a "temporary" fault by trying to close the breaker 2-3 times before accepting it as a "permanent" fault,
- Manual or automatic switching to reconfigure the power system in a timely manner by isolating the faulted section, then reconnecting the unfaulted sections,
- Device event logs,
- Digital fault recorders,

---

[34] Redundancy is multiple instances of the same software, firmware, devices, and/or data configured in an active/passive or load sharing mode. Redundancy for data and logs needs to be consistent with the organization's data retention plan and continuity of operations plan.

- Power quality (PQ) harmonics recorders, and
- Time synchronization to the appropriate accuracy and precision.

B.3 CENTRALIZED MONITORING AND CONTROL

- SCADA systems have approximately 99.98% availability with 24x7 monitoring,
- SCADA systems continuously monitor generators, substations, and feeder equipment (e.g., every second and/or report status and measurements "by exception"),
- SCADA systems perform remote control actions on generators, substations, and feeder equipment in response to operator commands or software application commands,
- Automatic Generation Control (AGC) issues control commands to generators to maintain frequency and other parameters within limits,
- Load Shedding commands can drop feeders, substations, or other large loads rapidly in case of emergencies,
- Load Control commands can "request" or command many smaller loads to turn off or cycle off,
- Disturbance analysis (rapid snapshots of power system during a disturbance for future analysis),
- Alarm processing, with categorization of high priority alarms, "intelligent" alarm processing to determine the true cause of the alarm, and events, and
- Comparisons of device settings against baseline settings.

B.4 CENTRALIZED POWER SYSTEM ANALYSIS AND CONTROL

Energy Management Systems (EMS) and Distribution Management Systems (DMS) use many software functions to analyze the real-time state and probable future state of the power system. These software functions include:

- "Power Flow" models of the transmission system, generators, and loads simulate the real-time or future (or past) power system scenarios,
- "Power Flow" models of the distribution system simulate real-time or future power system scenarios,
- State estimation uses redundant measurements from the field to "clean up" or estimate the real measurements from sometimes noisy, missing, or inaccurate sensor data,
- Power flow applications use the state estimated data to better simulate real-time conditions,
- Load and renewable generation forecasts based on weather, history, day-type, and other parameters forecast the generation requirements,
- Contingency Analysis (Security Analysis) assesses the power flow model for single points of failure (n-1) as well as any linked types of failures, and flags possible problems,

- Generation reserve capacity is available for instantaneous, short term, and longer term supply of generation in the event of the loss of generation,
- Ancillary services from bulk generation are available to handle both efficiency and emergency situations (e.g. generator is set to "follow load" for improved efficiency, generator is capable of a "black start" namely to start up during an outage without needing external power),
- Fault Location, Isolation, and Service Restoration (FLISR) analyze fault information in real-time to determine what feeder section to isolate and how to best restore power to unfaulted sections,
- Volt/VAR/Watt Optimization determine the optimal voltage, VAR, and generation levels usually for efficiency, but also to handle contingencies and emergency situations,
- Direct control of DER and loads (load management) for both efficiency and reliability,
- Indirect control of DER and loads (demand response) for both efficiency and reliability, and
- Ancillary services from DER for both efficiency and reliability (e.g., var support from inverters, managed charging rates for PEVs).

B.5 TESTING

- Lab and field testing of all power system and automation equipment minimizes failure rates,
- Software system factory, field, and availability testing,
- Rollback capability for database updates,
- Configuration testing,
- Relay coordination testing, and
- Communication network testing, including near power system faults.

B.6 TRAINING

- Dispatcher training simulator, using snapshots of real events as well as scenarios set up by trainers,
- Operational training using case studies, etc.,
- Training in using new technologies, and
- Security training.

B.7 EXAMPLE SECURITY TECHNOLOGY AND SERVICES

The selection and implementation of security technology and services is based on an organization's specification of security requirements and analysis of risk. This process is outside the scope of this report. Included below are some example security technologies and services that are provided as guidance. These are listed with some of the Smart Grid common technical requirements. The example security technologies and services for the unique technical requirements are included in the logical architectural diagrams included in this section.

Table B-2 Example Security Technologies and Services

Smart Grid Security Requirement	Smart Grid Requirement Name	Example Security Technologies/Services
SG.SC-15	Public Key Infrastructure Certificates	<ul><li>Cryptographic and key management support</li><li>Secure remote certificate enrollment protocol, with appropriate cert policies matching authorization policies</li></ul>
SG.SC-16	Mobile Code	<ul><li>Software quality assurance program ("the level of confidence that software is free from vulnerabilities, either intentionally designed into the software or accidentally inserted at anytime during its lifecycle and that the software functions in the intended manner."["National Information Assurance Glossary"; CNSS Instruction No. 4009 National Information Assurance Glossary])</li><li>Code inspection</li><li>Code-signing and verification on all mobile code</li><li>Allowed / Denied entities technology to detect mobile-code</li></ul>
SG.SC-18	System Connections	<ul><li>Identification and authorization</li><li>Information classification</li><li>Security domains and network segmentation</li><li>Allowed / Denied entities services</li><li>Allowed / Denied entities connections</li></ul>
SG.SC-19	Security Roles	<ul><li>Security management (data, attributes, functions, management roles, separation of duties)</li><li>Policy decision point (PDP) and Policy Enforcement Point (PEP) products</li><li>Role based access control (RBAC)</li><li>Training</li></ul>
SG.SC-20	Message Authenticity	<ul><li>Non-repudiation of origin</li><li>Non-repudiation of receipt</li><li>Message integrity</li></ul>
SG.SC-21	Secure Name/Address Resolution Service	<ul><li>Redundant name services</li><li>Restricting transaction entities based on IP address</li></ul>
SG.SC-22	Fail in Known State	<ul><li>Fail secure</li><li>Trusted recovery at the firmware and system levels</li><li>Software quality assurance program</li></ul>
SG.SC-30	Information System Partitioning	<ul><li>Traffic labeling and enforcement</li><li>Information classification program</li><li>Process (and Inter-process) access verification</li><li>Network-based and physical separation, labeling, etc.</li><li>RBAC technologies</li><li>Firewalls</li><li>OS-based process execution separation</li></ul>
SG.SI-8	Information Input	<ul><li>User data protection</li></ul>

Smart Grid Security Requirement	Smart Grid Requirement Name	Example Security Technologies/Services
	Validation	- Internal system data protection - RBAC - Separation of duties - Software quality assurance program - Internal system data protection - Non-repudiation - Authentication - Data transfer integrity - Before processing any input coming from a user, data source, component, or data service it should be validated for type, length, and/or range - Implement transaction signing - Access controls must check that users are allowed to use an action before performing the rendering or action
SG.SI-9	Error Handling	- Log management program - Delivery of error messages over secure channel - Software quality assurance program
SG.AC-6	Separation of Duties	- Security management (data, attributes, functions, management roles, separation of duties) - RBAC - Training
SG.AC-7	Least Privilege	- Security management (data, attributes, functions, management roles, separation of duties) - RBAC - Security domains and network segmentation - Traffic classification and priority routing
SG.AC-21	Passwords	- Authentication - Identification - Subject binding - Password Complexity Enforcement - Salted Hashes - Password Cracking Tests
SG.AC-9	System Use Notification	- System access history - Logon banner or message
SG.AC-8	Unsuccessful Login Attempts	- Authentication failure notice - Logon banner or message - Failed Login Attempt Lockouts
SG.AC-17	Access Control for Portable and Mobile Devices	- Limitation on scope of selectable attributes - Limitation on multiple concurrent sessions - System access banners - System access history - Limitation of network access - Secure communications tunnel - Authentication
SG.AC-16	Wireless Access	- Limitation on scope of selectable attributes

Smart Grid Security Requirement	Smart Grid Requirement Name	Example Security Technologies/Services
	Restrictions	- Limitation on multiple concurrent sessions - System access banners - System access history - Limitation of network access - Secure communications tunnel - Authentication
SG.AU-2	Auditable Events	- Event logging standard - Log management program - Scalable log filtering/parsing - Centralize logging/syslog to a NOC or SOC - 7x24 real-time auditing and automatic event notification
SG.AU-3	Content of Audit Records	- Event logging standard - Security audit event selection - Security audit review and analysis - Log management program - Scalable log filtering/parsing - Centralize logging/syslog to a NOC or SOC - 7x24 real-time auditing and automatic event notification
SG.AU-4	Audit Storage Capacity	- Record retention standards and requirements - Regular archiving and management of logs - Centralize logs to an enterprise log management system - Enable automatic file system checks for available disk space - Log management program
SG.AU-15	Audit Generation	- Security audit automatic response - Security audit automatic data generation - Verify that application level auditing is implemented in COTS and custom code - Verify that OS level auditing exists - Centralize logging/syslog to a NOC or SOC